Python Network Programming Techniques

50 real-world recipes to automate infrastructure networks and overcome networking challenges with Python

Marcel Neidinger

Packt>

BIRMINGHAM—MUMBAI

Python Network Programming Techniques

Group Product Manager: Vijin Boricha
Publishing Product Manager: Yogesh Deokar
Senior Editor: Sangeeta Purkayastha
Content Development Editor: Nihar Kapadia
Technical Editor: Shruthi Shetty
Copy Editor: Safis Editing
Project Coordinator: Shagun Saini
Proofreader: Safis Editing
Indexer: Rekha Nair
Production Designer: Nilesh Mohite

First published: September 2021
Production reference: 1120821

Published by Packt Publishing Ltd.
Livery Place
35 Livery Street
Birmingham
B3 2PB, UK.

ISBN 978-1-83864-663-9

`www.packt.com`

To my mom and dad – the most important writers in my life.

And to my colleague, Julio, without whom I wouldn't be where I am today.

Thank you for everything you have done for me.

Contributors

About the author

Marcel Neidinger started to program at the age of 10 and currently works as an API and programmability lead for the EMEAR Systems Engineering organization at Cisco Systems. Specifically, he works with customers and partners to build custom solutions using programmability and APIs.

Besides having a bachelor's degree in computer science, he is also a Cisco Certified DevNet Associate as well as a Cisco Certified DevNet Specialist for enterprise network automation.

You can find him on Twitter at `squ4rks` and on GitHub at `squ4rks`.

I want to thank my manager, Frans Wittenberg, and my director, Koen Jacobs, for enabling me to write this book. Thanks to Frederic Wagner for all the support inside and outside of work while writing this. Michael, Sven, Simon, Bernhard, Nikola, Christian, and Malte – thanks for bearing with me during this time.

About the reviewers

Wang Yin Parry, CCIE No. 40245, is a senior network engineer at **King Abdullah University of Science and Technology** (**KAUST**) in Saudi Arabia. Prior to working at KAUST, Parry left his footprints in the Royal Bank of Scotland, Apple, NCS, Equinix, and AT&T in Singapore. Parry has 12 years of experience in the computer networking industry and has certifications from Cisco, Microsoft, HPE, and Aruba. Parry is a NetDevOps evangelist in the Sinophone world; he was one of the first authors to publish Python tutorial books for network engineers in the Chinese-speaking world. His books *The Road to Python for Network Engineers* and *Manage IT Infrastructure Using Python* were respectively published in Mainland China and Taiwan in November 2020 and May 2021.

I started my journey in NetDevOps in my second year at KAUST in 2017. I could never have imagined it would be so successful in the years to come. I've been blessed with taking part in lots of great projects over the last few years at KAUST. I'd like to thank my team, in particular Khalid Mustafa, Kevin Sale, and Gary Corbett, for encouraging me and giving me the chance to participate in projects that give me the chance to apply what I have learned in the real world.

Rickard Körkkö is a NetDevOps consultant at SDNit, where he's part of a group of experienced technical specialists with a great interest in, and focus on, emerging network technologies.

He's a self-taught programmer with a primary focus on Python. His daily work includes working with orchestration tools such as Ansible to manage network devices.

He has also served as a technical reviewer for the book *Mastering Python Networking, Third Edition*, by *Eric Chou*.

Table of Contents

2
Connecting to Network Devices via SSH Using Paramiko

3
Building Configuration Templates Using Jinja2

4
Configuring Network Devices Using Netmiko

5

Model-Driven Programmability with NETCONF and ncclient

6

Automating Complex Multi-Vendor Networks with NAPALM

7

Automating Your Network Tests and Deployments with pyATS and Genie

8

Configuring Devices Using RESTCONF and requests

9

Consuming Controllers and High-Level Networking APIs with requests

10

Incorporating your Python Scripts into an Existing Workflow by Writing Custom Ansible Modules

11

Automating AWS Cloud Networking Infrastructure Using the AWS Python SDK

4

Configuring Network Devices Using Netmiko

5

Model-Driven Programmability with NETCONF and ncclient

6

Automating Complex Multi-Vendor Networks with NAPALM

7

Automating Your Network Tests and Deployments with pyATS and Genie

8

Configuring Devices Using RESTCONF and requests

9

Consuming Controllers and High-Level Networking APIs with requests

10

Incorporating your Python Scripts into an Existing Workflow by Writing Custom Ansible Modules

11

Automating AWS Cloud Networking Infrastructure Using the AWS Python SDK

12
Automating your Network Security Using Python and the Firepower APIs

Other Books You May Enjoy

Index

Preface

Networks and the connectivity they provide play a crucial role in the smooth operation of your business. However, with the rising pressure put on our networking infrastructure by ever-increasing traffic, paired with more and more devices sending and receiving traffic from more and more destinations both inside and outside of our networks, as well as the need to change configurations faster, more reliably, and possibly without any downtime and automation, becomes a central tool when tackling our modern infrastructure demands.

In this book, we will use Python to get hands-on experience of how to automate different aspects of our networks. With recipes, each chapter introduces you to one Python package used for network automation. From using a Python script to sending configuration commands to a network device via SSH to interacting with cloud-based REST APIs, each chapter introduces you to the tools needed to get started.

Who this book is for

Are you a network or infrastructure engineer who wants to get started using Python to automate your network? Then this book will help you to get started by teaching you, using hands-on examples, the programmability aspects and Python packages needed to complement your networking knowledge and get started with network automation.

What this book covers

Chapter 1, A Primer to Python 3, covers the basic concepts of Python used throughout this book. We will cover variables, loops, basic data structures, and flow control as well as concepts such as functions that help make scripts more readable.

Chapter 2, Connecting to Network Devices via SSH Using Paramiko, covers how to establish an SSH connection to our network device using Python and the paramiko package. We will then use this established connection to issue simple commands and read basic output back from the devices.

Chapter 3, Building Configuration Templates Using Jinja2, teaches you how to create templates for configuration files that can be filled with data provided by the Python script using Python and jinja2, a widely used templating language. Using loops, control structures, and inheritance within jinja2 templates, we will build highly customizable configuration templates.

Chapter 4, Configuring Network Devices Using Netmiko, looks at netmiko, which is built on top of paramiko. The netmiko package provides abstraction around SSH interactions with network devices of multiple vendors. Using netmiko, we will interact with the network device to retrieve status information and apply configuration changes.

Chapter 5, Model-Driven Programmability with NETCONF and ncclient, covers how model-driven programmability allows the user to specify the desired state of a networking infrastructure using a module that then gets applied to the device. We will use the `ncclient` package to change networking devices based on their YANG models.

Chapter 6, Automating Complex Multi-Vendor Networks with NAPALM, looks at NAPALM, which allows developers to automate complex multi-vendor networks using a unified interface. In this chapter, we will see examples of how to automate the configuration of devices from different vendors and validate these deployments.

Chapter 7, Automating Your Network Tests and Deployments with pyATS and Genie, looks at how automated network tests allow you to verify the success of your change quicker and find potential errors introduced by your change more easily. In this chapter, we will use pyATS and Genie to go from simple connectivity tests to advanced state tests for a variety of devices.

Chapter 8, Configuring Devices Using RESTCONF and requests, revisits model-driven programmability by using RESTCONF, a protocol that exposes a subset of NETCONF commands via HTTP(S). We will explore the `requests` package to send different RESTCONF JSON payloads to network devices.

Chapter 9, Consuming Controllers and High-Level Networking APIs with requests, discusses how most modern network controllers provide a REST API to make interaction with the underlying network infrastructure more uniform and easy to consume. In this chapter, we will learn how to query REST APIs using the `requests` package to retrieve information from and apply changes to multiple network devices.

Chapter 10, Incorporating Your Python Scripts into an Existing Workflow by Writing Custom Ansible Modules, covers Ansible, which is one of the most common tools for the automation of IT systems in the world. Its power comes from the variety of modules that provide functionality, and in this chapter, you'll learn how to write such a module yourself. This will allow you to call your Python scripts written in previous chapters from Ansible.

Chapter 11, Automating AWS Cloud Networking Infrastructure Using the AWS Python SDK, looks at **Amazon Web Services** (**AWS**), which is one of the biggest cloud providers in the world. In this chapter, we will use the AWS API and its Python library `boto3` to administer our cloud network.

Chapter 12, Automating Your Network Security Using Python and Firepower APIs, discusses how security is one of the biggest concerns in modern network engineering. In this chapter, we will have a look at how to automate our network security using the APIs available in Cisco's Firepower products and the `requests` module.

To get the most out of this book

This book requires you to have a version of Python, preferably Python 3, installed. All scripts have been tested against Python version 3.8 on Mac OS X and Linux but should also work on Windows.

Chapter 7, Automating Your Network Tests and Deployments with pyATS and Genie, and *Chapter 10, Incorporating Your Python Scripts into an Existing Workflow by Writing Custom Ansible Modules*, use Python packages that are incompatible with Windows. To follow along with these chapters, you will have to use either the Windows Subsystem for Linux or a virtual machine running Linux. Any modern Linux distribution, such as CentOS/Fedora or Debian/Ubuntu, should work.

In addition, you'll need a network device to test your scripts against. This device can be physical, virtual, or part of a sandbox, but an admin user is required.

Chapter 11, Automating AWS Cloud Networking Infrastructure Using the AWS Python SDK, requires a Firepower Management console from Cisco, and *Chapter 9, Consuming Controllers and High-Level Networking APIs with requests*, requires a Cisco Meraki organization. The code in both chapters can be tested using the sandboxes available with the Cisco DevNet sandbox offerings.

Software/hardware covered in the book	**OS requirements**
Python3	Windows, Mac OS X, or Linux (version >3.6)
Ansible	Linux or Mac OS X (any)

If you are using the digital version of this book, we advise you to type the code yourself or access the code via the GitHub repository (link available in the next section). Doing so will help you avoid any potential errors related to the copying and pasting of code.

Download the example code files

You can download the example code files for this book from GitHub at `https://github.com/PacktPublishing/Python-Networking-Cookbook`. In case there's an update to the code, it will be updated on the existing GitHub repository.

We also have other code bundles from our rich catalog of books and videos available at `https://github.com/PacktPublishing/`. Check them out!

Code in Action

Code in Action videos for this book can be viewed at `https://bit.ly/3s93enF`

Download the color images

We also provide a PDF file that has color images of the screenshots/diagrams used in this book. You can download it here: `http://www.packtpub.com/sites/default/files/downloads/9781838646639_ColorImages.pdf`.

Conventions used

There are a number of text conventions used throughout this book.

`Code in text`: Indicates code words in text, database table names, folder names, filenames, file extensions, pathnames, dummy URLs, user input, and Twitter handles. Here is an example: "Mount the downloaded `WebStorm-10*.dmg` disk image file as another disk in your system."

A block of code is set as follows:

```
import requests
from requests.auth import HTTPBasicAuth

auth = HTTPBasicAuth("root", "test")
```

When we wish to draw your attention to a particular part of a code block, the relevant lines or items are set in bold:

```
my_hostname = "router01"
```

Any command-line input or output is written as follows:

```
HTTP/1.1 200 OK
Content-Type: application/json; charset=utf-8

{
    "id": 25,
    "name": "my device",
    "mac": "01-23-45-67-89-AB-CD-EF"
}
```

Bold: Indicates a new term, an important word, or words that you see onscreen. For example, words in menus or dialog boxes appear in the text like this. Here is an example: "On the security credentials page, navigate down to the **Access Keys** section and click on **Create New Access Key** to generate a new key pair."

> **Tips or important notes**
> Appear like this.

Get in touch

Feedback from our readers is always welcome.

General feedback: If you have questions about any aspect of this book, mention the book title in the subject of your message and email us at `customercare@packtpub.com`.

Errata: Although we have taken every care to ensure the accuracy of our content, mistakes do happen. If you have found a mistake in this book, we would be grateful if you would report this to us. Please visit `www.packtpub.com/support/errata`, selecting your book, clicking on the Errata Submission Form link, and entering the details.

Piracy: If you come across any illegal copies of our works in any form on the Internet, we would be grateful if you would provide us with the location address or website name. Please contact us at `copyright@packt.com` with a link to the material.

If you are interested in becoming an author: If there is a topic that you have expertise in and you are interested in either writing or contributing to a book, please visit `authors.packtpub.com`.

Reviews

Please leave a review. Once you have read and used this book, why not leave a review on the site that you purchased it from? Potential readers can then see and use your unbiased opinion to make purchase decisions, we at Packt can understand what you think about our products, and our authors can see your feedback on their book. Thank you!

For more information about Packt, please visit `packt.com`.

Share Your Thoughts

Once you've read *Python Network Programming Techniques*, we'd love to hear your thoughts! Scan the QR code below to go straight to the Amazon review page for this book and share your feedback.

`https://packt.link/r/1838646639`

Your review is important to us and the tech community and will help us make sure we're delivering excellent quality content.

1
A Primer on Python 3

Since its development in the early 1990s, Python has become one of the most widely used programming languages in the world. Its easy-to-use yet powerful syntax and cross-platform availability, combined with a rich ecosystem of pre-built functionality, have turned it into one of the most widely used languages in the realm of infrastructure automation. From simple automation scripts to complex web applications, Python can handle a wide variety of tasks.

In this chapter, we are going to review some basic concepts of the Python programming language before using these skills in the following chapters to automate our infrastructure.

While not being a full introduction to the Python language, this chapter should serve as a refresher for the concepts we are going to use in the automation script in later chapters. If you are already proficient with Python, you can skip ahead to the next chapter.

Specifically, you will learn about the following concepts in this chapter:

- Assigning variables in Python
- Converting between data types in Python
- Looping over lists in Python
- Controlling the flow of a Python program using `if` statements

- Executing code until a condition is met using `while` loops
- Writing reusable code with functions
- Storing and accessing key-value pairs using dictionaries
- Importing modules from the standard library
- Installing modules from the **Python Package Index** (**PyPI**)

Technical requirements

For this section and for the remainder of the book, you'll need an installation of Python—specifically, you'll need a Python interpreter of version 3.6.1 or higher. This book makes use of language constructs of Python 3 and thus is incompatible with Python 2.x.

You will also require a code editor. Popular choices include Microsoft **Visual Studio Code** (**VS Code**) or Notepad++.

While the code for this book was written on a Mac with macOS X version 10.15 installed, we don't use any packages that behave differently on Windows or Linux, and all code samples should behave the same.

You can view this chapter's code in action here: `https://bit.ly/3fW3EsK`

Assigning variables in Python

Variables are named placeholders we can use to store and manipulate data. Variables point to a portion of the memory in which the data is stored. The kind of data that is stored within the memory portion that a variable is pointing to is known as the data type of that variable. These data types can be, for example, a string for text or an integer for a whole number. While some programming languages (such as C++ or Java) require you as the developer to explicitly define what kind of data you want to put into a variable, Python is dynamically typed. This means that the programming language will take care of assigning the correct data type to your variable, based on what you are putting in.

In this recipe, you will see how to assign variables in Python using the = operator and see the basic data types Python offers.

Getting ready

Open your code editor and start by creating a file called `variables.py`. Next, navigate your terminal to the same directory in which you just created the `variables.py` file.

How to do it...

Let's start by assigning some variables containing text (also referred to as a string), whole and floating-point numbers (in the form of integers and floats), truth values (known as Booleans), and lists that hold a list of variables:

1. Define a variable called `my_hostname` and assign it the value of your name using the = operator:

```
my_hostname = "router01"
```

2. We can also use single quotes to assign a string. Note that you can't mix the different types of quotes:

```
my_domain = 'example.com'
```

3. Furthermore, we can use triple quotes to define a string that spans multiple lines:

```
my_motd = """
This is a string that will
contain linebreaks and could be the motd of a
router.
"""
```

4. If we want to see the value of a variable, we can use Python's built-in `print()` function to display the value stored on the command line:

```
print(my_hostname)
print(my_domain)
print(my_motd)
```

5. Similar to strings, we can also assign numeric values. We can distinguish two types of variables: floats for floating-point numbers and integers for whole numbers. Define a variable called `my_port` to hold the default port number of a **Secure Shell** (**SSH**) server (as an integer/whole number) and a variable called `my_throughput` to hold your server's average number of requests per minute (as a float/floating-point number):

```
my_port = 22
my_throughput = 1.75
```

6. We can print out the value of a variable of the type `int` or `float`, the same way we did with the string:

```
print(my_port)
print(my_throughput)
```

7. Truth values in Python are represented by the two `True` and `False` Boolean values. We can define a variable called `knows_python`, assign it to either be true or false, and then print the value of said variable:

```
knows_python = True
knows_python = False
print(knows_python)
```

8. Sometimes, we want to store lists (also referred to as arrays or vectors in other programming languages) of values or variables. First, we define our variable to be an empty list:

```
l = []
```

9. Next, we want to add variables to the end of this list. Let's add `my_hostname`, `my_domain`, `my_port`, and `my_throughput` variables to the list using the `append()` function:

```
l.append(my_hostname)
l.append(my_domain)
l.append(my_port)
l.append(my_throughput)
```

10. We can access elements in our list using the index the element is saved under. Be aware that indexes start at zero, so the first element in the list will be at index `0`, not at index `1`. Lists preserve the order in which items are added, so the first element in our list will be the first name, the second element will be the last name, and the third element will be the age. Let's print out this information:

```
print("The hostname stored in the list is: ")
print(l[0])
print("The domain stored in the list is: ")
print(l[1])
print("The port stored in the list is: ")
print(l[2])
```

11. Lastly, if you want to check what kind of data type a variable has, you can use the `type()` function. Let's use the `type()` function to first print out the data type of our list and then the data type of the element at index `0`:

```
print(type(l))
print(type(l[0]))
```

12. To run the Python script we just created, open your terminal and execute the script by typing the `python3 variables.py` command. Your output should look like this:

```
python3 variables.py
router01
example.com

This is a string that will
contain linebreaks and could be the motd of a
router.

False
The hostname stored in the list is:
router01
The domain stored in the list is:
example.com
The port stored in the list is:
22
<class 'list'>
<class 'str'>
```

How it works...

Python determines the type of variable based on the value you are providing it with. The main data types you are going to be dealing with are strings for text, integers for whole numbers, floats for floating-point numbers, Booleans for truth values, and lists for lists of variables and values.

The type of number, float, or integer is determined based on whether you are assigning a number with or without a decimal point.

As you can see from the preceding example, where we are adding variables of the string type and variables of the integer and float type to the list, lists are not restricted to only storing one data type, but rather can keep variables of all different kinds.

When dealing with lists, it is important to remember that the index of a list starts at zero, so the first element of a list is not at index `1` but rather at index `0`.

When trying to determine which data type a variable has, the `type()` function can be used to return this information back to the user, as seen in *Step 11* in the preceding code sample.

Converting between data types in Python

While Python does determine the data type of a variable automatically, we often like to explicitly convert a variable from one data type to another. An example of where this is useful could be a user providing an input—for example, their age—from the terminal. The variable would have the `str` data type, but we might need it as an integer. This is where converting between data types—or type casting—comes into play.

In this recipe, you will see how to receive an input from a user and convert the received string into an integer.

Getting ready

Open your code editor and start by creating a file called `convert_variables.py`. Next, navigate your terminal to the same directory in which you just created the `convert_variables.py` file.

How to do it...

In this recipe, we will receive the user's age as input in the form of a string. We will then convert that string into a number, add `10` to it, and convert it back to a string:

1. First, we need to capture the input from the user. To do this, we can use the built-in `input()` function:

```
the_age = input("What's your age?: ")
```

2. Next, we convert the input from a string to an integer and add `10` to it:

```
age_as_num = int(the_age)
age_as_num = age_as_num + 10
```

3. Lastly, we convert the variable we just added `10` to into a string and print that string back to the user:

```
age_as_text = str(age_as_num)
print("Your age + 10 is " + age_as_text)
```

4. So far, we have always explicitly converted the data types. When, as in this example, we want to have our variables of the `int` type included (or formatted) into a string, we can also use the f-string syntax:

```
print(f"Your age + 10 is {age_as_num}")
```

5. To run this script, go to your terminal and execute it by typing the command `python3 convert_variables.py`. Your output should look like the following example. The age that is calculated will, of course, depend on the input you are giving:

```
python3 convert_variables.py
What's your age?: 26
Your age + 10 is 36
Your age + 10 is 36
```

How it works...

The conversion of variables between data types works by invoking the correct conversion function. In this example, we used `int()` to convert the string our user provided into an integer, and later used `str()` to convert the integer back to a string.

The different conversion functions for the basic data types we have encountered so far are outlined here:

- `str()` to convert to strings
- `int()` to convert to integers
- `bool()` to convert to a Boolean—for example, `bool(0)` will convert to `False` while `bool(1)` converts to `True`
- `float()` to convert to a floating-point number

If we had not done the last conversion (from number to string), Python would have errored out as it does not know how to add a string to an integer. This error is referred to as a `TypeError` error and looks like this:

```
TypeError: can only concatenate str (not "int") to str
```

As you can see, this error message is pretty detailed as to what went wrong. `TypeError` specifies that there was an error related to types, and the following text specifies that we were trying to concatenate a variable whose data type is integer to a variable whose data type is string. However, Python only knows how to concatenate a string with a string, not how to concatenate an integer with a string.

To avoid these errors, it is thus good practice to always cast variables you plan on using in strings (that is, for printing them back to the user) to strings explicitly.

Formatted string literals, commonly referred to as f-strings, allow us to create a shorthand method for this quite common task of piecing together strings from variables of different data types. By referencing the name of the variable (in our case, `age_as_num`) in curly braces, Python will substitute this with the value of the variable and will automatically call the string conversion function on it.

Looping over lists in Python

A common use case when programming is the need to apply the same kind of operation to a list of variables. Think, for example, of a script that needs to connect to a few routers. For this to work, the script will need to know the **Internet Protocol** (**IP**) addresses of said routers and then perform the same set of commands on them. Instead of copying and pasting the same code multiple times and just changing the router IP, we can create a list of our IPs and then execute the commands.

In this recipe, you will see how to create a list of IP addresses and print them out using a `for` loop. We will reuse the same concept in the next chapter to actually connect to the device and issue configuration commands.

Getting ready

Open your code editor and start by creating a file called `loops.py`. Next, navigate your terminal to the same directory in which you just created the `loops.py` file.

How to do it...

In this recipe, we will create **IP version 4** (**IPv4**) addresses with a static network part and a changing host part that is retrieved from a list:

1. First, we need to define the network part of our IP address:

```
network_part = "192.168.2."
```

2. Next, we need to define the host parts for each of our three routers. We will have routers on `192.168.2.20`, `192.168.2.40`, and `192.168.2.60`:

```
host_parts = [20, 40, 60]
```

3. Now, we can use a `for-each` loop to iterate over each item in the list and combine the network and host part into a valid IP address. Since the host parts are integers, we will also need to convert them:

```
for host_part in host_parts:
    ip = network_part + str(host_part)
    print("The router IP is: " + ip)
```

4. To run this script, go to your terminal and execute the `python3 loops.py` command. Your output should look like this:

```
python3 loops.py
The router IP is: 192.168.2.20
The router IP is: 192.168.2.40
The router IP is: 192.168.2.60
```

How it works...

In this example, we used a `for` loop to iterate over a given list of variables. As you can see, there is an indention that shows Python which parts of the code are to be executed in the loop. Contrary to other languages, where the indention is good to have but not necessary, Python uses indention to structure the program. The two indented lines that get executed for every variable in the `host_parts` list are also called the loop body.

There's more...

In this example, we used a `for each` loop to quickly iterate over all items in the list. Instead, we could have also used the length of the list and the index to retrieve the correct item. The following code would be equivalent to what we have seen in the preceding example:

```
network_part = "192.168.2."
host_parts = [20, 40, 60]

for idx in range(len(host_parts)):
    host_part = host_parts[idx]
```

```
    ip = network_part + str(host_part)
    print("The router IP is: " + ip)
```

In this example, we used two new functions. The `len()` function returns the length of a list. In this example, the `len()` function would return `3`. The `range()` function then returns a list of integers, starting from the first function argument (`0`, in this case) up to—but excluding—the second function argument (the length of the list, which is `3` in this case). So, in this example, `range(0, len(host_parts))` would return the list `[0, 1, 2]`, which are the three indices we can use to access all elements in our list.

Controlling the flow of a Python program using if statements

So far, we have only written scripts that ran each of the instructions linearly one after another. But quite often, we want to change the behavior of our script based on the value of a variable. Let's say, for example, that we want to write a script that asks the user what kind of device—a user device or an infrastructure device—they want to configure. Our script then returns an IP address with a pre-defined network part and a user-defined host part.

In this recipe, you will see how to create such a basic IP address management tool `if` clause.

Getting ready

Open your code editor and start by creating a file called `if_conditions.py`. Next, navigate your terminal to the same directory in which you just created the `if_conditions.py` file.

How to do it...

In this recipe, we will create a list of user and infrastructure devices, check what kind of device the user wants to retrieve, and then retrieve an IP from the correct list:

1. First, we need to define the two network parts for infrastructure and user devices:

```
infra_network_part = "10.2.10."
user_network_part = "10.20.1."
```

2. Next, read the user input to understand what kind of device we want to configure and which host part the user wants:

```
device_type = input("What type? [infra/user] ")
host_part = input("What's the host part? ")
```

3. We can now use an `if` clause to react, based on the choice of the user. We can react to the two different outcomes we know of—the user wanting to configure an infrastructure or a user device—and a third outcome, which is everything else. If the user has entered neither `user` nor `infra`, we can just exit the program with a message stating that we don't know that type:

```
if device_type == "infra":
        ip = str(infra_network_part) + str(host_part)
        print(f"Your infra ip: {ip}")
elif device_type == "user":
        ip = str(user_network_part) + str(host_part)
        print(f"Your user ip: {ip}")
else:
        print("Sorry, I don't know how to handle the type
{device_type}")
```

4. To run this script, go to your terminal and execute it by typing the command `python3 if_conditions.py`. Your output should look like the following example. The exact output depends on whether you choose to get the IP address of a user or of an infrastructure device, and on which host part you specified:

```
python3 if_conditions.py
What type? [infra/user] infra
What's the host part? 30
Your infra ip: 10.2.10.30
```

How it works...

In this example, we used an `if` clause to determine which network part to use based on the user input. Similar to loops, the part of the program that should be executed if the condition is met—in our case, the check whether the `device_type` variable is equal to `infra` or `user`—is indented.

In this example, we checked for two explicit conditions: the `device_type` either being `infra` or `user`, and a catch-all with an `else` statement. This part of the conditional is executed when none of the preceding explicit conditions are met.

Note how we used the double equals sign to check which value the variable has. Since the single equals sign is already reserved for assigning variables, we have to use the double equals sign when checking our variables in `if` clauses.

There's more...

The double equals sign is an expression in Python that returns a Boolean. We can check this by creating a new script with the following code:

```
print(type(10 == 10))
```

In this example, we are comparing whether `10` is equal to `10`, which it is. More interestingly, the data type that is revealed by the `type()` function is `<class 'bool'>`, which is the data type of a Boolean in Python.

In the preceding `if` clause, we checked for equivalency, but sometimes we want to check that a variable is not equal to or greater than another variable or value. Python offers a multitude of comparison operators.

The following list shows you all comparison operators for two variables, `a` and `b`, that are available to you:

- `a == b` checks whether the value of `a` is equal to the value of `b`.
- `a != b` checks whether the value of `a` is not equal to the value of `b`.
- `a > b` checks whether the value of `a` is strictly greater than the value of `b`.
- `a < b` checks whether the value of `a` is strictly less than the value of `b`.
- `a >= b` checks whether the value of `a` is greater than or equal to the value of `b`.
- `a <= b` checks whether the value of `a` is less than or equal to the value of `b`.

Executing code until a condition is met using while loops

A very common use case for command-line programs is to keep asking a user for more information until a certain condition is met. With our current set of tools, this would be difficult. We know how to ask a user for information (using the `input()` function) and we know how to check information (using `if` clauses), but how could we keep executing a piece of code until a condition (that is, the correct input from a user) is met? We could use a `for` loop with an extremely high number, but this number would also eventually come to an end.

To address this, Python includes a `while` loop. `while` loops are essentially a combination of a loop and an `if` clause that allow us to execute a piece of code until a condition is met.

In this recipe, you'll see how to use a `while` loop to keep asking a user for more IP addresses that we will add to a list. Once the user is satisfied with the addresses they have added, we will terminate the loop and print out the list of addresses.

Getting ready

Open your code editor and start by creating a file called `while_loops.py`. Next, navigate your terminal to the same directory in which you just created the `while_loops.py` file.

How to do it...

In this recipe, we will create a script that keeps asking the user for IP addresses and later prints a complete list back:

1. First, we need to create a list that will hold our addresses, as well as a variable that will contain our user input:

```
ips = []
user_in = ""
```

2. Next, we can define our `while` loop. Everything in this loop's body will be executed until the specified condition is met:

```
while user_in != "done":
    user_in = input("IP addressor type done to exit: ")
    if user_in != "done":
        ips.append(user_in)
```

3. Finally, once the user has exited the loop by typing `done` in the terminal, we want to print our list of IPs back:

```
print(f"Your IPs are: {ips}")
```

4. To run this script, go to your terminal and execute it by typing the command `python3 while_loops.py`. Your output should look like the following example. The exact output depends on which and how many IP addresses you have entered:

```
python3 while_loops.py
IP address or type done to exit: 192.158.0.1
IP address or type done to exit: done
Your IPs are: ['192.158.0.1']
```

How it works...

In this example, we used a `while` loop to execute a piece of code that asks for a user input. This script will run until the user has met the condition of the `while` loop by providing `done` as the input—in other words, the code within the `while` block will keep executing as long as the condition is `True`.

Once the condition is met, the code after the `while` loop begins executing. In this case, the code is just printing back the list of IPs we have just added.

Be aware that `while` loops can become infinite loops. These are loops that never exit, and thus your script never exits. Infinite loops occur when the conditional in the `while` loop can never be `False`.

There's more...

The infinite loops mentioned as a problem previously might also be desirable in some cases. Consider a web server, for example. This server has a loop that is running forever, listening for new incoming web requests that will then be handled.

We can create such infinite loops by just specifying `True` as the condition in the `while` loop. `True` will, as the name states, never evaluate to `False`, and thus the loop will keep running:

```
while True:
    print("I am executing.")
```

But could we also break out of such an infinite loop or are we, once started, stuck in this loop forever?

Python offers us the ability to escape from `while` loops (and `for` loops as well) by using the `break` keyword:

```
while True:
    best_num = input("What's your favorite number? ")
    if best_num == "42":
        break
    else:
        print(f"{best_num} is not the best number. Try 
again.")
```

In this example, we ask the user for input on which is the best number. We then check that variable and, if the number provided by the user is equal to `42`, we break out of the `while` loop using the `break` keyword.

Writing reusable code with functions

So far, all the code we have written has been executed directly. But what if we want to reuse some part of our code—maybe a workflow to check whether an IP address matches your infrastructure or user IP ranges—in another part of our script? We could copy and paste the code that we wrote, but what if some part of our workflow—for example, the network parts of our infrastructure devices—changes? We would have to change that information in every part of our script we copied the workflow to! This is not only labor-intensive but also very error-prone.

In this recipe, you will see how to create your own functions that you can use to make your workflows reusable.

Getting ready

Open your code editor and start by creating a file called `functions.py`. Next, navigate your terminal to the same directory in which you just created the `functions.py` file.

How to do it...

In this recipe, we will create a function that checks—based on a provided network and host part—whether an IP address is an infrastructure or a user device:

1. First, we need two lists containing the network parts for infrastructure and user devices:

```
infra_parts = ["10.2.10.", "10.30.2."]
user_parts = ["10.50.2.", "10.60."]
```

2. Next, we need to define a function called `check_ip`, which we are going to put our workflow into. `check_ip` has one argument: the IP address we want to check.

   In the function itself, we loop through the two lists and check whether the IP starts with any of those prefixes. If it does, we will print the corresponding address type back to the user:

```
def check_ip(ip_addr):
    for net_part in infra_parts:
        if ip_addr.startswith(net_part):
            print(f"{ip_addr} is of type 'infra'")
    for net_part in user_parts:
        if ip_addr.startswith(net_part):
            print(f"{ip_addr} is of type 'user'")
```

3. Lastly, we need to test our function by calling it with two test IPs:

```
check_ip("10.2.10.1")
check_ip("10.50.2.3")
```

4. To run this script, go to your terminal and execute it by typing the command `python3 functions.py`. Your output should look like this:

```
python3 functions.py
10.2.10.1 is of type 'infra'
10.50.2.3 is of type 'user'
```

How it works...

In this example, we defined a new function called `check_ip()`. The function received one argument called `ip_addr` and then used `startswith()`, a built-in function of strings in Python, to check whether the `ip_addr` argument provided to the function starts with any of the prefixes we defined in our lists before.

There's more...

In this simple example, we only had a single argument provided to the function. However, a function can have multiple arguments. The following is a simple function that takes two arguments, casts them into integers, adds them up, and then returns the result to the user:

```
def add_nums(num_a, num_b):
    num_a = int(num_a)
    num_b = int(num_b)

    result = num_a + num_b
    print(f"The result of {num_a} + {num_b} is {result}")
add_nums(10, 2)
```

Speaking of returning, we can also have our function return back the result of its computation instead of printing it out. This can be useful when we want to use the output of one function in the computation of another.

In the following example, we use a `return` statement to return the result instead of directly printing it:

```
def add_num_and_return(num_a, num_b):
        num_a = int(num_a)
        num_b = int(num_b)

        result = num_a + num_b

        return result

res = add_num_and_return(10, 12)
print("The result is {res}")
```

Storing and accessing key-value pairs using dictionaries

One of the most common tasks when automating or programming is the storage of key-value pairs. Think, for example, of a configuration file or a user profile. These are long text files containing key-value pairs. So, we need a data structure in Python to accommodate this kind of variable.

In this recipe, you will see how to store configuration information in a dictionary, manipulate the values of said configuration, and print the values of our configuration afterward.

Getting ready

Open your code editor and start by creating a file called `dictionaries.py` and a file called `config.txt`. Next, navigate your terminal to the same directory in which you just created the `dictionaries.py` file.

How to do it...

In this recipe, we will read a configuration file from the hard drive, create a dictionary to store the configuration information we retrieved from the file, and allow the user to query and change configuration variables. At the end, we overwrite the original configuration file with the changes provided by the user.

1. First, we need to create our initial configuration file. Open the `config.txt` file you created and add the following dummy configuration:

```
mode=development
host=192.168.20.1
port=8020
user=admin
password=password
```

2. We now need to read this file into our program. Python already has a function to read and write files built-in. Open up the `dictionaries.py` file you created and put the instructions for *Step 2* to *Step 7* in the previously opened `dictionaries.py` file:

```
config = {}
with open("config.txt", "r") as f:
lines = f.readlines()
```

3. We can now loop over each line of our configuration file, split the line to key and value, and then add it to our `config` dictionary:

```
for line in lines:
            key, value = line.split("=")
            value = value.replace("\n", "")
            config[key] = value
            print(f"Added key {key} with value {value}")
```

4. Next, we can ask the user for a key to retrieve. The dictionary has a handy function to check whether a key is in the `dictionaries` list of keys:

```
user_key = input("Which key would you like to see? ")
if user_key not in config:
    print(f"I don't know the key {user_key}")
# We will add code here in step 6
```

5. Now that we know that the key the user has chosen is actually in our configuration, we can retrieve it and print it back to the user, and ask whether the user wants to change the value of the key. To do so, replace the `# We will add code here in step 6` comment from *Step 4* with the following code:

```
else:
        val = config[user_key]
        print(f"Current value for {user_key}:{val}")
        next_step = input("Would you like to change?[y/n]")
```

6. If the user says yes (or `y`) to our question of changing the value, we can overwrite the value in our dictionary with something the user provides:

```
if next_step == "y":
        new_val = input("What is the new value? ")
        config[user_key] = new_val
```

7. So far, we have only changed the value of the configuration, but we have not yet changed the `config.txt` file. To do so, we can iterate over all key-value pairs in our dictionary and write them to the file. To do so, we will open the file again, but this time in `write` mode:

```
        with open("config.txt", "w") as f:
                for key, value in config.items():
                        l = f"{key}={value}\n"
                                f.write(l)
```

8. To run this script, go to your terminal and execute it by typing the command `python3 dictionaries.py`. Your output should look like the following example, varying only depending on which configuration variable you choose to change:

```
python3 dictionaries.py
Added key mode with value development
Added key host with value 192.168.20.1
Added key port with value 8020
Added key user with value admin
Added key password with value password
Which key would you like to see? user
Current value for user:admin
Would you like to change it? [y/n]y
What's the new value? root
```

How it works...

In this example, we read the contents of a file into our program and stored it in a dictionary.

To define a dictionary in Python, simply assign a variable an empty dictionary with the two curly braces.

When reading a file in Python, we first need to open it. The `open()` function requires a path or name of the file, and optionally takes a mode. In our example, we used the `w` mode, for *write*. This opens the file with read/write access. If we only need to read a file, we can use the `r` mode.

We used the `open()` function together with a `with` block. When dealing with files, we need to close them. We could do this explicitly by calling the `f.close()` function, but the `with` block will do this for us once all instructions within the block have been executed.

The `split()` function takes a string and splits it according to the provided delimiter, and then returns a list. In our example, this list always has the length `2`, the key (the part that was before the = sign in `config.txt`) will be at index `0`, and the value (the part that came after the = sign in `config.txt`) will be at index `1`.

We can then retrieve the key the user wants and check whether it is present in our dictionary using an `if` clause and the `in` operator we saw already when iterating over a list of items.

If the key the user has provided is, in fact, part of our `config` dictionary, we can retrieve and later overwrite its value by using the key in square brackets, similar to how we used the indices to retrieve the entry of a list.

There's more...

When opening a file, we have more than just the two `r` and `w` modes available to us. The following is a list of available modes:

- `r+` opens a file for reading and writing.
- `w+` opens a file for reading and writing, creating a file if it does not exist already. Writing starts at the beginning of the file.
- `a` opens a file for writing. Any lines that are written to this file will be appended to the end of it.
- `a+` opens a file for reading and writing. Any lines that are written to this file will be appended to the end of it.

In this example, we have chosen to explicitly open the file for reading and re-open it for writing. In production code, you would have most likely wanted to open the file once at the beginning and used the `w+` mode.

In the example, we used the `.items()` function to retrieve key-value pairs from our dictionary. While this is the most common use case, we can also retrieve a list of just the keys and just the items. The following example shows both functions:

```
d = {
    "name": "Marcel",
    "height": 1.73
}
for k in d.keys():
    print(f"A key is {k}")
for v in d.values():
    print(f"A value is {v}")
```

Importing modules from the standard library

Python is often described as *batteries-included* programming. This stems from the fact that Python comes with an extensive standard library that provides a lot of functionality already. While some functions such as `print()` or `input()` are loaded by default, a lot of the functionality can be imported into our program. This code, also referred to as a module, encapsulates the functionality and can be used without the need to install anything, as it comes pre-packaged with your Python installation.

Getting ready

Open your code editor and start by creating a file called `libs.py`. Next, navigate your terminal to the same directory in which you just created the `libs.py` file.

How to do it...

In this recipe, we will import the `random` module. This module allows us to generate random numbers or draw a random element from a list. We will use the list of lower- and uppercase letters, as well as digits provided by the `string` module to generate a password:

1. First, we need to import the required modules using the `import` statement:

```
import random
import string
```

2. We first need to know how many random uppercase characters, lowercase characters, and digits our password should contain. For this, we can use the `random.randint()` function. This function, when called with the arguments `1` and `7`, will return a random integer number between `1` and `7`. This random number will determine how many uppercase letters, lowercase letters, and digits our password will get:

```
num_upper = random.randint(1, 7)
print(f"Adding {num_upper} uppercase letters to the
password")
num_lower = random.randint(1, 7)
print(f"Adding {num_lower} lowercase letters to the
password")
num_digits = random.randint(1, 7)
print(f"Adding {num_digits} digits to the password")
```

3. With the random numbers for each of our character categories ready, we can go ahead and generate our password. We could write a list of uppercase characters, lowercase characters, and digits ourselves, but we can also use the ones provided in the `strings` module. Since we are carrying out the same functionality, retrieving a variable number of elements from a list and appending them to a string, let's write a function so that we do not have to copy and paste our code when we want to reuse it:

```
def add_to_password(num_to_include, choices):
    temp = ""
    for i in range(num_to_include):
        temp = temp + str(random.choice(choices))
    return temp
uppers = add_to_password(num_upper, string.ascii_
uppercase)
lowers = add_to_password(num_lower, string.ascii_
lowercase)
numbers = add_to_password(num_digits, string.digits)

pwd = uppers + lowers + numbers
print(f"Your password is {pwd}")
```

4. To run this script, go to your terminal and execute it by typing the command `python3 libs.py`. The output should look like the following example. The number of uppercase letters, lowercase letters, and digits will, of course, be different, as well as the resulting password:

```
python3 libs.py
Adding 5 uppercase letters to the password
Adding 6 lowercase letters to the password
Adding 7 digits to the password
Your password is PYJFFcdfjub7316218
```

How it works...

We used the `import` statement to signal to our Python interpreter to make the functionality in this module available to use in our script. After importing the module, we were able to use the functions it provides by simply calling the name of the function, prefixed by the name of the module.

There's more...

In this example, we included the entire module and then called the functions in the module by always prefixing the function name with the name of the module. While this always works, it can be cumbersome, especially when using the function a module provides many times in our code, or if the module name is long.

Instead of importing the entire module, we can also just import a single function that we can then call without prefixing it with the module name. Here is the code to import the `randint()` function from the `random` module:

```
from random import randint

random_num = randint(0, 10)
print(f"Your random number is {random_num}")
```

Installing modules from the PyPI

While the Python standard library provides an extensive set of functionalities, more specific use cases—such as connecting to a networking device to issue configuration commands—are not covered specifically. While we *could* implement them using the functionality provided in the standard library, this work has already been done by the community that contributed packages back into the Python ecosystem, more specifically by making them available for you to install from PyPI. Each chapter of this book will cover one such package that is specific for network automation.

In this recipe, we will use the `pip` package manager (which is a recursive acronym for **pip installs packages**) to install the `paramiko` package we will use in the next chapter.

When installing packages from PyPI, we are installing third-party modules. Contrary to the modules that come with the standard library, these third-party modules need to be downloaded from the internet, and thus having internet access is required.

Getting ready

Open up your terminal.

How to do it...

In this recipe, we will call `pip` to install a package. We will first install the latest version of a package, uninstall it again, and then see how to install a specific version of a package:

1. Install the latest version of `paramiko` by typing the following command in your terminal. This command will install `paramiko` and all of its dependencies for you:

```
python3 -m pip install paramiko
```

2. To uninstall a package, you can type the following command. It will ask you for confirmation to remove the package:

```
python3 -m pip uninstall paramiko
```

3. Often, it is useful to install the specific version of a package you have tested your code against instead of always installing the latest version. To do so, `pip` offers you a way of specifying the version. The following command will install version 2.7.1 of `paramiko`:

```
python3 -m pip install paramiko==2.7.1
```

4. You can get a list of all packages that are installed on your package by using the `freeze` command. The returned list will include the version number for each module so that you can verify you have installed the correct version:

```
python3 -m pip freeze
```

5. When publishing code on GitHub or other code-sharing sites, many project authors include a list of packages that need to be installed in a text file. This file is usually called `requirements.txt`, and you can install all the packages listed in it using the following command:

```
python3 -m pip install -r requirements.txt
```

How it works...

When issuing the `pip install` command, `pip` will reach out to PyPI to find a package that matches the name you provided. Depending on whether you specified a version or not, it will download and install either the version you specified or the latest version. `pip` also resolves, downloads, and installs all packages that the package you are trying to install depends on.

Requirements files have made it easy to install all the dependencies of a project. While this file could be called whatever you want, it has become convention to call it `requirements.txt`. You can create such a `requirements.txt` file by issuing the following command in your command line:

```
python3 -m pip freeze > requirements.txt
```

2
Connecting to Network Devices via SSH Using Paramiko

When administrating IT devices from a remote location, be it network equipment or servers, SSH has become the standard. With its secure transport and various authentication methods, it's a safe choice that is widely used to this day to administer and configure servers or network devices. It is thus only natural when getting started with programmability and network automation, to find a way of issuing SSH commands not by hand but from a script. With this, you can take a sequence of commands you used to type into the device by hand and execute them programmatically on one or more devices. The last part is crucial. With a script that executes commands *for you*, you can easily apply the same sequence of commands to another device.

While we could implement the SSH protocol ourselves, this would be cumbersome work. Luckily, the Python community has already developed an SSH client library that is available for our use, called Paramiko.

In this chapter, we are going to learn the basics of programmatically connecting to a network device using SSH. We are going to use Cisco devices for our examples but the workflow is the same regardless of the vendor.

In this chapter, we will work through the following recipes:

- Initiating an SSH session with Paramiko
- Executing a command via SSH
- Reading the output of an executed command
- Executing the same command against multiple devices
- Executing a sequence of commands
- Using public/private keys for authentication
- Loading local SSH configuration

Technical requirements

For this section and the remainder of the book, you'll need an installation of Python. Specifically, you'll need a Python interpreter of version 3.6.1 or higher. This book makes use of language constructs of Python 3 and thus is incompatible with Python 2.x. If you haven't done so already in the previous chapter, please go ahead and install the Paramiko package (`python3 -m pip install paramiko`). At the time of writing, we are using the latest version of Paramiko, version 2.7.1. You may install this exact version by issuing `python3 -m pip install paramiko==2.7.1`.

You also want a code editor. Popular choices include Microsoft Visual Studio Code or Notepad++. Additionally, you'll need a device (virtual or physical) that you can log into via SSH.

You can view this chapter's code in action here: `https://bit.ly/37Ih46N`

Initiating an SSH session with Paramiko

The basis of connecting to a device via SSH with Python and Paramiko is the `SSHClient` object of the library. We will use this object to create an initial connection to the SSH server and later we will use the functions of this object to execute commands on the device.

In this recipe, you will see how to programmatically open an SSH connection.

Getting ready

Open your code editor and start by creating a file called `initiating.py`. Next, navigate your terminal to the same directory that you just created the `initiating.py` file in.

How to do it...

Let's start by importing the Paramiko library and creating a client object. We will also specify the host, username, and password in variables and then initiate a connection to the specified host:

1. Import the Paramiko library:

```
from paramiko.client import SSHClient
```

2. Specify the host, username, and password. You can name these variables however you like. In the Python community, it has become standard to uppercase these global variables. The three variables `SSH_USER`, `SSH_PASSWORD`, and `SSH_HOST` are variables of type string and we thus use double quotes to mark them. The `SSH_PORT` variable is an integer and thus does not use double quotes:

```
SSH_USER = "<Insert your ssh user here>"
SSH_PASSWORD = "<Insert your ssh password here>"
SSH_HOST = "<Insert the IP/host of your device/server
here>"
SSH_PORT = 22 # Change this if your SSH port is different
```

3. Create an `SSHClient` object, which we just imported from Paramiko:

```
client = SSHClient()
```

4. While we have created our client object, we have not yet connected to the device. We will use the `connect` method of the client object to do so. Before actually connecting, we will need to make sure that our client knows the host keys:

```
client.load_system_host_keys()
try:
    client.connect(SSH_HOST, port=SSH_PORT,
                            username=SSH_USER,
                            password=SSH_PASSWORD,
                            look_for_keys=False)
    print("Connected successfully!")
except Exception:
    print("Failed to establish connection.")
```

5. Finally, we have created our connection. It is a good habit to close connections after we are done using them. To do so, we can use the `close()` function of our client:

```
finally:
    client.close()
```

6. To run this script, go to your terminal and execute it with the following:

```
python3 initiating.py
```

How it works...

In this example, we first imported the Paramiko library's `SSHClient` class. Next, we set up our connection details. While you could also provide these details directly when calling the `connect` method on the client object, it is good practice to put them into variables. This means that if you are creating multiple client objects at different points of your script, you don't have to change the username/password and host in each of these calls but just once in the variables. Be careful when submitting these scripts to your colleagues or uploading them to code hosting services though, as they do contain your login details. You can have a look at the *There's more* section of this recipe to see how you can either prompt for this information interactively or get it from your environment.

Before connecting to the device, using the `connect` method, we load the host keys. SSH uses host keys and the fingerprints of an SSH server to make sure that the IP you are connecting to is the server you connected to before. When connecting to a brand-new device, you'll have to accept this new host key. While we can keep a separate set of host keys for our Paramiko client, it is usually best to just use the host keys that the user executing the script has. By doing this, you can connect to every device you have previously connected to from your command line, also from your Paramiko scripts.

To do this loading, we are using the `load_system_host_keys()` function. This function searches the default known hosts file used by OpenSSH and copies them over into Paramiko. See the following section for an example of how to make Paramiko accept new keys by default.

With our host keys configured, we can now actually connect to the device that we have specified. To do so, we are using a `try-catch` block. This is a Python construct that allows us to `catch` exceptions, errors that can be raised by any part of the code we are using, and handle them. Python will attempt to execute the instructions in the `try` block. If any part of that code, in our example the `connect()` method, errors out and raises an exception, Python will jump into the `except` block and execute the instructions within that block. In our example, we are only printing out a message that our connection was unsuccessful. The third block, our `finally` block, will be executed both when the connection has been successful (our `except` block was not executed) as well as after a failed connection (our `except` block was executed). This allows us to clean up after both a successful and unsuccessful connection and avoids dangling SSH connections.

There's more...

In this example, we relied on the user having already logged into the device from their command line in order for the host to be known. If we use the preceding code to connect to a device that was not previously known, the code will fail with an exception.

The way Paramiko handles unknown host keys can be specified using a policy. One of these policies, `AutoAddPolicy`, allows us to just add unknown host keys to our scripts set of host keys:

```
from paramiko.client import SSHClient, AutoAddPolicy
SSH_USER = "<Insert your ssh user here>"
SSH_PASSWORD = "<Insert your ssh password here>"
SSH_HOST = "<Insert the IP/host of your device/server here>"
SSH_PORT = 22 # Change this if your SSH port is different

client = SSHClient()
client.set_missing_host_key_policy(AutoAddPolicy())
client.connect(SSH_HOST, port=SSH_PORT,
                            username=SSH_USER,
                            password=SSH_PASSWORD)
```

The preceding code will automatically add these host keys. Be aware that this might be a potential security risk since you are not verifying that the host you are connecting to is the one you connected to last time.

In this example, we passed connection details such as username, hostname, and password directly as a variable in the script. While this is great for testing, you might want to have your script prompt you for a variable upon execution. For non-secret variables such as the username, host, and port, we can use the built-in `input()` function, but for passwords, it's better to use a dedicated password prompt that hides what you have typed so that someone looking over your console history can't retrieve your password. For this purpose, Python has the built-in `getpass` module.

Follow these steps to retrieve the configuration variables necessary, not as static information in the script but rather interactively from the user using a combination of input and the `getpass` module:

```
import getpass
SSH_PASSWORD = getpass.getpass(prompt='Password: ',
stream=None)
SSH_USER = input("Username: ")
SSH_HOST = input("Host: ")
SSH_PORT = int(input("Port: "))
```

Executing a command via SSH

With our connection now open, we can go ahead and execute a command on our remote device. Similar to the way we deal with issuing commands on a remote device by hand, we have three different streams that come back to us: the standard out (or `stdout`), which is the normal output, the standard error (or `stderr`), which is the default stream for the system to return errors on, and the standard in (or `stdin`), which is the stream used to send text back into the executed command. This can be useful if, in your workflow, you would normally interact with the command line.

In this recipe, you will see how to programmatically open an SSH connection and then send a command of your choice to the device.

Getting ready

Open your code editor and start by creating a file called `command.py`. Next, navigate your terminal to the same directory that you just created the `command.py` file in.

How to do it...

Let's start by importing the Paramiko library and create a client object as seen in the last recipe. We'll then execute a single command of your choice on this device:

1. Import the Paramiko library:

```
from paramiko.client import SSHClient
```

2. Specify the host, username, and password. You can name these variables however you like. In the Python community, it has become a standard to uppercase these global variables:

```
SSH_USER = "<Insert your ssh user here>"
SSH_PASSWORD = "<Insert your ssh password here>"
SSH_HOST = "<Insert the IP/host of your device/server
here>"
SSH_PORT = 22 # Change this if your SSH port is different
```

3. Create an `SSHClient` object, which we just imported from Paramiko:

```
client = SSHClient()
```

4. While we have created our client object, we have not yet connected to the device. We will use the `connect` method of the client object to do so. Before actually connecting, we will need to make sure that our client knows the host keys:

```
client.load_system_host_keys()
client.connect(SSH_HOST, port=SSH_PORT,
                         username=SSH_USER,
                         password=SSH_PASSWORD)
```

5. Finally, we can use the client to execute a command. Executing a command will return three different file-like objects to us representing `stdin`, `stdout`, and `stderr`:

```
CMD = "show ip interface brief" # You can issue any
command you want
stdin, stdout, stderr = client.exec_command(CMD)
client.close()
```

6. To run this script, go to your terminal and execute it with this:

```
python3 command.py
```

How it works...

In this example, we first created a new client as seen in the previous example. We then used the `exec_command()` method to execute a command of our choice.

The function returns three different file-like objects for the three different streams: `stdin`, `stdout`, and `stderr`. In the next recipe, *Reading the output of an executed command*, we will use this to read back the output that was provided when executing a command.

Reading the output of an executed command

In the previous recipe, we saw how to first connect to a device and then execute a command. So far, we have ignored the output though.

In this recipe, you will see how to programmatically open an SSH connection, send a command, and then write the output of that command back to a file. We will use this to back up a running configuration.

Getting ready

Open your code editor and start by creating a file called `read_out.py`. Next, navigate your terminal to the same directory that you just created the `read_out.py` file in.

How to do it...

Let's start by importing the Paramiko library and create a client object as seen in the last recipe. Then execute a single command of your choice on this device and save the output:

1. Import the Paramiko library:

```
from paramiko.client import SSHClient
```

2. Specify the host, username, and password. You can name these variables however you like. In the Python community, it has become a standard to uppercase these global variables:

```
SSH_USER = "<Insert your ssh user here>"
SSH_PASSWORD = "<Insert your ssh password here>"
SSH_HOST = "<Insert the IP/host of your device/server
here>"
SSH_PORT = 22 # Change this if your SSH port is different
```

3. Create an `SSHClient` object, which we just imported from Paramiko:

```
client = SSHClient()
```

4. While we have created our client object, we have not yet connected to the device. We will use the `connect` method of the client object to do so. Before actually connecting, we will need to make sure that our client knows the host keys:

```
client.load_system_host_keys()
client.connect(SSH_HOST, port=SSH_PORT,
                         username=SSH_USER,
                         password=SSH_PASSWORD)
```

5. Finally, we can use the client to execute a command. Executing a command will return three different file-like objects to us representing `stdin`, `stdout`, and `stderr`:

```
CMD = "show running-config"
stdin, stdout, stderr = client.exec_command(CMD)
```

6. We will use the `stdout` object to retrieve what the command has returned:

```
output = stdout.readlines()
```

7. Next, we write the output back to a file:

```
with open("backup.txt", "w") as out_file
    for line in output:
        out_file.write(line)
```

8. To run this script, go to your terminal and execute it with this:

```
python3 read_out.py
```

How it works...

In this example, we first created a new client as seen in the previous example. We then used the `exec_command()` method to execute a command of our choice. We then used the returned file-like object for the standard out to read everything that our remote device printed into the standard out and write it into a file.

A file-like object in the context of Python is an object that offers the same functions as a file object. When using the `open()` function to create a new file object to read/write from locally, we are offered some functions such as `write()` or `readlines()`. These functions allow us to read or write from or to a file. A file-like object offers the same functions and can thus be used as if it was a remote file. We use the `readlines()` function on the `stdout` object to read everything returned – in this example, the running configuration of our device – and write it line by line to a local file.

There's more...

If you prefer to just see the output of your command instead of writing it to a file, you can also use the following construct, instead of the code provided in *step 7* of the recipe, to print out your entire configuration:

```
for line in output:
    print(line.strip())
```

The `strip()` method used in the preceding example will delete any unnecessary leading or trailing whitespaces.

Executing the same command against multiple devices

In the previous recipes, we have always only dealt with a single device. Quite often we have a fleet of similar devices that we want to configure in unison.

In this recipe, you will see how to programmatically open an SSH connection to multiple devices, issue the same command to all of them, and then save the output. We will again use this example to back up the running configuration of multiple devices.

Getting ready

Open your code editor and start by creating a file called `exec_multiple.py`. Next, navigate your terminal to the same directory that you just created the `exec_multiple.py` file in. Additionally, create a file called `credentials.json`. We will use this file to retrieve credentials such as the username and password of our devices.

How to do it...

Let's start by creating our credentials file. We will then read that file from our Python script, create clients for each of these devices, and finally execute a command while also saving the output back to our file:

1. Import the required libraries, Paramiko and `json`:

```
import json
from paramiko.client import SSHClient
```

2. Open up the `credentials.json` file and provide the credentials to your device(s) in the format shown in the following code. You can specify as many devices as you want:

```
[
 {
   "name": "<insert a unique name of your device>",
   "host": "<insert the host of your device>",
   "username": "<insert the username>",
   "password": "<insert the password",
   "port": 22
 },
 {
   "name": "<insert a unique name of your device>",
   "host": "<insert the host of your device>",
   "username": "<insert the username>",
   "password": "<insert the password",
   "port": 22
 }
]
```

3. Go back to your `exec_multiple.py` file. We will now open the JSON file in our Python script:

```
credentials = {}
with open("credentials.json") as fh:
     json.load(fh)
```

4. Create a variable holding the command you want to execute. We will then loop over all the devices specified in our `credentials.json` file and create an SSH client object. Additionally, we will create an individual output file for each of our devices based on the name we specified in the JSON file:

```
CMD = "show running-config"
for cred in credentials:
    out_file_name = str(cred['name']) + ".txt"
    client = SSHClient()
    client.load_system_host_keys()
    client.connect(SSH_HOST, port=cred['port'],
                             username=cred['username'],
                             password=cred['password'])
    stdin, stdout, stderr = client.exec_command(CMD)

    out_file = open(out_file_name, "w")
    output = stdout.readlines()
    for line in output:
        out_file.write(line)
    out_file.close()
    client.close()
    print("Executed command on " + cred['name'])
```

5. To run this script, go to your terminal and execute it with this:

```
python3 exec_multiple.py
```

How it works...

In this example, we first create a JSON file containing the credentials and connection details for all of our devices. We can view this file as a type of inventory. In general, it is good practice to keep this information separate from the Python code that is acting upon it.

We then use the built-in `json` module to read the credentials file into a list of dictionaries that we can loop over.

Based on the credentials specified in the credentials file, we then open up a new Paramiko connection, execute the command, and write the output to a new log file, that is, by including the name we set for our device in the JSON file, which is unique for each of the devices.

With this, we can back up the running configuration of an entire fleet of devices from one simple script.

Executing a sequence of commands

In the previous recipes, we have always only dealt with a single command that we wanted to execute. Maybe you have tried adding another call to the `exec_command()` function to the client already and have run into an error telling you that the session is closed. It is indeed correct that, once your command is done executing, the connection will close.

But quite often we don't want to execute only one but a sequence of commands one after another. We could reconnect for each of the commands, but this is a workaround that would create a lot of unneeded disconnecting and reconnecting. What we can do instead is, if the target device's configuration allows it, open up a shell session. This shell session is a single command, and we can then use `stdin`, `stderr`, and `stdout` to send multiple commands in the same session.

In this recipe, you will see how to programmatically open an SSH connection to a device, open a shell, and then send a list of commands to the device before closing the connection.

Getting ready

Open your code editor and start by creating a file called `exec_multiple_commands.py`. Next, navigate your terminal to the same directory that you just created the `exec_multiple_commands.py` file in.

How to do it...

Let's start by creating our credentials file. We will then read that file from our Python script, create clients for each of these devices, and finally execute a command while also saving the output back to our file:

1. Import the Paramiko library. We will also need the built-in `time` library:

```
from paramiko.client import SSHClient
import time
```

2. Specify the host, username, and password. You can name these variables however you like. In the Python community, it has become a standard to uppercase these global variables:

```
SSH_USER = "<Insert your ssh user here>"
SSH_PASSWORD = "<Insert your ssh password here>"
SSH_HOST = "<Insert the IP/host of your device/server
here>"
SSH_PORT = 22 # Change this if your SSH port is different
```

3. Create an `SSHClient` object, which we just imported from Paramiko:

```
client = SSHClient()
```

4. While we have created our client object, we have not yet connected to the device. We will use the `connect` method of the client object to do so. Before actually connecting, we will need to make sure that our client knows the host keys:

```
client.load_system_host_keys()
client.connect(SSH_HOST, port=SSH_PORT,
                         username=SSH_USER,
                         password=SSH_PASSWORD)
```

5. Open up an interactive shell session and a channel that we can use to retrieve the output:

```
channel = client.get_transport().open_session()
shell = channel.invoke_shell()
```

6. Next, specify the list of commands we want to execute on the device:

```
commands = [
     "configure terminal",
     "hostname test"
]
```

7. Iterate over each of the commands, execute them, and then wait for 2 seconds:

```
for cmd in commands:
     shell.send(cmd + "\n")
      out = shell.recv(1024)
      print(out)
     time.sleep(1)
```

8. Finally, we need to close the connection:

```
client.close()
```

9. To run this script, go to your terminal and execute it with this:

```
python3 exec_multiple_commands.py
```

How it works...

In this example, we use Paramiko's concept of an interactive shell to send multiple commands to the remote device one after another.

When invoking a shell with Paramiko, we will retrieve a channel. This channel was used by Paramiko in the background all along to execute our commands but has been hidden from us so far. The channel takes care of low-level aspects such as sending and receiving the data to and from the raw network connection to our device. The `channel` function we use in this example is the `send()` function, which sends a string to the remote device. Mind the carriage return we added to the command. The same way you, when connecting to a device via SSH, have to *execute* a command by typing `enter`, the interactive session has to indicate that via a linebreak, which is the same symbol sent by your interactive session when hitting the *Enter* key.

We are using the `sleep` function here to wait for the commands you have passed to have finished executing. For short-running commands, you can get away with not using this `sleep` function.

Using public/private keys for authentication

So far, we have always used a username/password combination to connect to our device. This is not the most secure way, however, and many security policies advocate using public-private key pairs instead of a static password.

In this recipe, you will see how to programmatically open an SSH connection using a password-protected private key.

Getting ready

Open your code editor and start by creating a file called `key_file.py`. Next, navigate your terminal to the same directory that you just created the `key_file.py` file in.

You'll also need a password-protected private key for the device/server you are trying to connect to and have the device/server configured to allow or require key-based logins.

How to do it...

Let's start by importing the required libraries, define our new connection details, and finally open up a connection using key-based authentication:

1. Import the Paramiko library:

```
from paramiko.client import SSHClient
```

2. Specify the host and username. You can name these variables however you like. In the Python community, it has become a standard to uppercase these global variables. Instead of the device password, we will now need two new variables – the path to the private key file that we want to use to authenticate and the password for that private key file:

```
SSH_USER = "<Insert your ssh user here>"
SSH_HOST = "<Insert the IP/host of your device/server
here>"
SSH_PORT = 22 # Change this if your SSH port is different
SSH_KEY = "<Insert the name of your private key here>"
SSH_KEY_PASSWORD = "<Insert the password here>"
```

3. Create an `SSHClient` object, which we just imported from Paramiko:

```
client = SSHClient()
```

4. While we have created our client object, we have not yet connected to the device. We will use the `connect` method of the client object to do so. Before actually connecting, we will still need to make sure that our client knows the host keys:

```
client.load_system_host_keys()
client.connect(SSH_HOST, port=SSH_PORT,
                         username=SSH_USER,
                         look_for_keys=True,
                         key_filename=SSH_KEY,
                         passphrase=SSH_KEY_PASSWORD)
```

5. As seen before, we can now execute a command once the connection is established:

```
stdin, stdout, stderr = client.exec_command('<your
command>')
```

6. Finally, we need to close the connection:

```
client.close()
```

7. To run this script, go to your terminal and execute it with this:

```
python3 key_file.py
```

How it works...

In this example, we use Paramiko's ability to load RSA keys for authentication to avoid using a username/password combination for authentication.

We need to import the same packages we have used before. The difference lies in the parameters that we pass to the `connect` function. Instead of specifying a password, we specify the name of our `ssh` key. The library will then, as indicated by setting the `look_for_keys` flag to true, search common places such as `~/.ssh` for keys and match them with the name provided. The `passphrase` argument is used to provide the passphrase used to decode the private key. If your private key does not have a passphrase, you can omit this argument.

Once the connection is established, we can use the client in the same way as we did before, when dealing with username-password authentication.

There's more...

In the preceding example, we relied on the key file being present in one of the known paths. Sometimes you might want to explicitly specify the path you are loading a key file from. You can do so by, instead of just specifying the filename in the `key_filename` attribute, specifying the entire path where Paramiko can find your private key.

For example, if your private key is in `/home/user/my_keys/id_rsa`, you could modify the preceding example like so:

```
SSH_KEY = "/home/user/my_keys/id_rsa"
```

If you want to connect to different devices and have multiple keys, one for each device, you can also pass a list of key names or paths to `ssh` keys to the `key_filename` attribute:

```
SSH_KEY = [
        "/home/user/my_keys/device_1",
        "/home/user/my_keys/device_2",
        "/home/user/my_keys/device_3"
]
```

Paramiko will then try out all the keys in the provided list for each device you are connecting to.

Loading local SSH configuration

When dealing with multiple different devices and connecting to them via SSH, it can be convenient to specify information such as the hostname, port, username, or identity file to use in a specific configuration file. The OpenSSH implementation stores this file in a file called config in the `.ssh` directory in your home directory (`~/.ssh/config` on macOS and Linux).

While we could copy and paste this information into our Python scripts or try to write a parsing function for the format ourselves, it is easier and more convenient to use Paramiko's SSH configuration parser.

In this recipe, you will see how to programmatically parse your `SSHConfig` file, extract the relevant information based on a host, and store it in a dictionary.

Getting ready

Open your code editor and start by creating a file called `parse_config.py`. Next, navigate your terminal to the same directory that you just created the `parse_config.py` file in.

You'll also need an SSH config file for the device you are trying to connect to. In this example, we will be using a config file that has the following content:

```
Host example
    Host <insert your host address here>
    User <insert your user here>
    Port <insert the port here>
    IdentityFile <insert the path to your private key here>
```

How to do it...

Let's start by importing the required libraries and defining the path to our SSH configuration:

1. Import the Paramiko library:

```
from paramiko.client import SSHClient
from paramiko import SSHConfig
```

2. Specify the path to your SSH config file and the name of your host as it appears in your SSH configuration (example in this snippet). We will populate all the other variables from the configuration we are reading:

```
SSH_CONFIG = "<insert path to ssh config here>"
SSH_HOST = "example"
```

3. Create an `SSHConfig` object, which we just imported from Paramiko, and create a local file object with the path to our SSH configuration:

```
config = SSHConfig()
config_file = open(SSH_CONFIG)
```

4. Next, we need to tell the `SSHConfig` object to load and parse the configuration file:

```
config.parse(config_file)
```

5. With the config parsed, we can now do a lookup on this configuration object to extract all information stored in the configuration itself. The lookup function will return a dictionary:

```
dev_config = config.lookup(SSH_HOST)
```

6. With our device configuration extracted from the SSH config we can go ahead and fill our connection details with what we have extracted from the SSH configuration file:

```
client.load_system_host_keys()
HOST = dev_config['hostname'],
client.connect(HOST, port=int(dev_config['port']),
                     username=dev_config['user'],
                     key_filename=dev_
config['identityfile'])
```

7. With the connection established, we can do all the different things we discovered in previous recipes before finally closing the connection:

```
client.close()
```

8. To run this script, go to your terminal and execute it with this:

```
python3 parse_config.py
```

How it works...

In this example, we use Paramiko's ability to parse an SSH configuration file to not define this information in multiple different locations.

We start by importing the `SSHConfig` class, in addition to the already established `SSHClient` class. Instead of manually specifying the host, username, and key file information, we now create a local file object that points to our SSH configuration.

With that file opened, we can now have Paramiko parse this configuration. The `SSHConfig` object now contains all the different information for each of the hosts. We can then do a lookup on our host – in this recipe, the host is called `example` – and extract all configuration variables that are known in the configuration file.

From that, we proceed to providing that information to the `SSHClient`. Instead of statically specifying it, we just access it from the dictionary that was returned by Paramiko when doing the lookup on the host.

3

Building Configuration Templates Using Jinja2

Configurations are the lifeblood of networking devices and could be, to draw a relation to software engineering, regarded as the *source code* of a network. When writing such configuration files, it can be handy to build templates that parts can be reused from. In this chapter, we will see how we can programmatically build and render such templates using a templating language called Jinja2.

The applications of this are manifold. Think of a script that reaches out to your **IP Address Management** (**IPAM**) system and automatically builds a configuration based on the information and the template you have written. Using sub-templates, you can significantly improve the overall cohesiveness of your configurations. Instead of each network engineer writing their own take on some type of configuration, you can pull in a template that does exactly this configuration for you.

While we could build our configuration templates with pure Python, it is more convenient to use a templating language. Templating languages are small languages that offer features such as loops or `if`-clauses so that you can implement rudimentary logic directly within your template.

Specifically, we will use Jinja2 for our templates and learn about the following applications of Jinja2:

- Loading Jinja2 templates in Python
- Passing variables from Python to a template
- Writing your rendered template to a file
- Using `for`-loops in Jinja2 to configure an access list
- Creating a port configuration template using `if`-clauses in Jinja2
- Creating modular templates using Jinja2 import methods
- Using Python functions from within your template with Jinja2 filters
- Structuring your configuration template with blocks and inheritance

Technical requirements

For this section and the remainder of the book, you'll need an installation of Python. Specifically, you'll need a Python interpreter of version 3.6.1 or higher. This book makes use of language constructs of Python 3 and thus is incompatible with Python 2.x. Additionally, you'll need to install the Jinja2 package. You can install the newest version of Jinja2 using `python3 -m pip install Jinja2`. At the time of writing, the current version is version 2.11.2.

You also want a code editor. Popular choices include Microsoft Visual Studio Code or Notepad++. Additionally, you'll need a device (virtual or physical) that you can log in to via SSH.

You can view this chapter's code in action here: `https://bit.ly/3CF7RdP`

Loading Jinja2 templates in Python

In this chapter, we will use our Python scripts as both an information source for our template and as the software that starts the rendering of our template into a string that we can then either display or write to a file for later usage.

In this recipe, we will set up the directory structure for our templates and create a Jinja2 environment in our Python script. This Jinja2 environment will then allow us to load and render our first template, written in Jinja2, from Python.

Getting ready

Open your code editor and start by creating a file called `render_template.py`. Next, navigate your terminal to the same directory that you just created the `render_template.py` file in.

Next, in the same directory as your Python file, create a directory called `templates`. Inside of this directory, create a file called `first.conf.tpl`.

How to do it...

Let's start by writing our Python script to load and render our first template before adding some content into the template:

1. Import the required classes from the Jinja2 library:

```
from Jinja2 import Environment, FileSystemLoader
```

2. We need to tell Jinja2 where to load our templates from. In this chapter, we will always load from the filesystem:

```
loader = FileSystemLoader("templates")
```

3. Next, we can create our Jinja2 environment. This environment is used to obtain a template object that we can then later render out:

```
environment = Environment(loader=loader)
```

4. With the environment defined, we can obtain our template:

```
tpl = environment.get_template("first.conf.tpl")
```

5. Finally, we can render our template, save the output to a variable, and print that variable out on the command line:

```
out = tpl.render()
print(out)
```

6. With the Python part done, let's open the `first.conf.tpl` file. This is the template that is being rendered. For now, just put some normal text into the file:

```
Hello from the Jinja2 template!
```

7. To run this script, go to your terminal and execute it with the following:

```
python3 render_template.py
```

How it works...

In this recipe, we first import our required Jinja2 components, `Environment` and `FileSystemLoader`, into our script.

`FileSystemLoader` specifies where Jinja2 can find our templates. While you could put them wherever you want as long as you specify the correct path here in the argument, it has become common practice to create a subfolder called `templates` and store all your templates in there.

With the loader created, we can now initiate the Jinja2 environment. This environment handles the retrieval of template objects for us and it is from this environment that we can obtain our template object itself by using the `get_template()` method.

Note that, when using the `get_template()` method, you don't have to specify the full path to your template. The environment will automatically search in the directory that you specified in `FileSystemLoader`.

With our template retrieved from the system, we can then go ahead and render it using the `render()` method. This method will return the rendered version of the template as a string that we can then print out.

Our template called `first.conf.tpl` is just a plain text file. Again, you can call this file whatever you want, but it has become common to append `.template` or `.tpl` at the end of the file name to signify that this file is a template.

In the coming recipes, you will see how we can incorporate logic and other advanced features directly into our template.

There's more...

Some code editors struggle with proper syntax highlighting when a file has the extension `tpl` or `template`. If you prefer to have syntax highlighting for your configuration file, you can also call your file `<filename>.template.<extension>`, that is, `first.template.conf`.

This will preserve the syntax highlighting of your editor.

Passing variables from Python to a template

So far, we have only seen a static template and we have not passed any information from our Python script back into the template.

The power of this concept, however, is that we can pass information that we have obtained in our Python script, for example, by querying an API or by parsing the output of an SSH command we issued on a device, and use that information in our template.

Luckily, Jinja2 supports an easy mechanism for passing variables from our script into our template, so in this recipe, we will see how we can do this.

Getting ready

Open your code editor and start by creating a file called `vars.py`. Next, navigate your terminal to the same directory that you just created the `vars.py` file in.

In the same directory as your Python file, create a directory called `templates` if you have not already done so in a previous recipe. Inside of this directory, create a file called `vars.conf.tpl`.

How to do it...

Let's start by writing our Python script to load and render our template before passing a variable into it and printing the rendered result back to us:

1. Import the required classes from the Jinja2 library:

```
from Jinja2 import Environment, FileSystemLoader
```

2. We need to tell Jinja2 where to load our templates from. In this chapter, we will always load from the filesystem:

```
loader = FileSystemLoader("templates")
```

3. Next, we can create our Jinja2 environment. This environment is used to obtain a template object that we can then later render out:

```
environment = Environment(loader=loader)
```

4. With the environment defined, we can obtain our template:

```
tpl = environment.get_template("vars.conf.tpl")
```

5. Finally, we can render our template. The `render` method allows us to pass an arbitrary number of variables into the template by specifying them as arguments of the `render()` method. In this example, we are going to ask for the name of the person running the script and pass that information on to the template:

```
user_name = input("What's your name?: ")
out = tpl.render(name=user_name)
print(out)
```

6. With the Python part done, let's open the `vars.conf.tpl` file. This is the template that is being rendered. Here, we are going to use the variable we have just passed into the `render` method:

```
Hello {{ name }}
```

7. To run this script, go to your terminal and execute it with the following:

```
python3 vars.py
```

How it works...

With our Jinja2 components such as the environment and filesystem loader loaded and initialized, we can go ahead and retrieve some information in the Python script that can later be passed into the template.

In this example, we used the `input` method to ask the user for some information and then passed the variable on to the `render` method. You can pass as many variables as you want into the template. The name that you give the function argument, in this case, `name`, will be the new name of this variable when using it within the template.

In our template file, we can now use this variable by enclosing the name with double curly brackets. This indicates to Jinja2 that the content of this bracket represents a variable whose value should be put in the place of it.

There's more...

Be aware that Jinja2 fails silently. This means that variables you call in your template that were not passed will just show up as empty. See the following example where we pass the `user_name` variable but use the `user` variable in our template.

In our script, type the following code snippet:

```
out = tpl.render(user_name=name)
```

In our template file, use the following code snippet:

```
Hello {{ user }}
```

The result of rendering this template would be the following:

```
Hello
```

Writing your rendered template to a file

So far, we have always printed our template back to the user via the command line, and while this can be nice to debug and check whether our script ran correctly, in most cases, we want to save our rendered template as a configuration file to our filesystem.

In this recipe, we will see how we can write the same template we rendered in the *Passing variables from Python to the template* recipe to our filesystem using Jinja2 template streams.

Getting ready

Open your code editor and start by creating a file called `save_to_file.py`. Next, navigate your terminal to the same directory that you just created the `save_to_file.py` file in.

Next, in the same directory as your Python file, create a directory called `templates` if it does not already exist from a previous recipe. Inside of this directory, create a file called `vars.conf.tpl` if it does not already exist from a previous recipe.

How to do it

Let's start by writing our Python script to load and render our template before passing a variable into it. Instead of rendering the template and printing the result back to us, we will dump it directly into a local file:

1. Set up the Jinja2 environment. For a detailed explanation of the steps, please refer to the *Loading Jinja2 templates in Python* recipe in this chapter:

```
from Jinja2 import Environment, FileSystemLoader
loader = FileSystemLoader("templates")
environment = Environment(loader=loader)
```

2. With the environment defined, we can obtain our template:

```
tpl = environment.get_template("vars.conf.tpl")
```

3. Next, let's ask the user for the name again:

```
user_name = input("What's your name?: ")
```

4. Instead of rendering the stream, we are going to create a template stream and dump that into a file called `rendered.conf`:

```
tpl.stream(name=user_name).dump("rendered.conf")
```

5. With the Python part done, let's open the `vars.conf.tpl` file. This is the template that is being rendered. Here, we are going to use the variable we have just passed into the `render` method:

```
Hello {{ name }}
```

6. To run this script, go to your terminal and execute it with the following:

```
python3 save_to_file.py
```

How it works...

With our Jinja2 components, such as the environment and filesystem loader loaded and initialized, and the information necessary to render the template retrieved from the user, we can go ahead and write the rendered output of our template to a file.

Contrary to the preceding examples, we used the `stream()` method instead of the `render()` method. This method creates an object that works like a Python generator. The template is parsed and every time a new line is rendered, this line is returned by the stream.

While we could also iterate over this stream with a `for`-loop, Jinja2 conveniently includes a `dump()` method. This method takes the name of the file we want to write the output of our template to as an argument (in our case, `rendered.conf`) and dumps all the lines returned from our rendered template into this file.

Using for-loops in Jinja2 to configure an access list

While we can already achieve a lot by being able to pass information from our Python file into a template and use this information in there, this is something that could be done with standard Python strings.

The true power of Jinja2 lies in its ability to include logic in the template. Basically, Jinja2 templates can also be seen as programs written in the Jinja2 template language.

This language supports such constructs as `for`-loops and `if`-clauses, which we have already seen in Python. This means that instead of copy-pasting the same text over and over and only changing one small part of it, we can create loops to make this more compact, readable, and changeable.

In this recipe, you'll see how to take two lists of IP addresses that are provided in your Python file and create an ACL for all the IPs in both lists, denying access for the IPs in one list while allowing access for IPs in the other list.

Getting ready

Open your code editor and start by creating a file called `jinja_loops.py`. Next, navigate your terminal to the same directory that you just created the `jinja_loops.py` file in.

Next, in the same directory as your Python file, create a directory called `templates` if it does not already exist from a previous recipe. Inside of this directory, create a file called `acl.conf.tpl`.

How to do it...

Let's start by setting up our Python script to be able to render a template. In our script, we will then define the two lists holding those IPs that will be denied access and those that will be allowed through.

We will then print the result back to the user:

1. Set up the Jinja2 environment. For a detailed explanation of the steps, please refer to the *Loading Jinja2 templates in Python* recipe in this chapter:

```
from Jinja2 import Environment, FileSystemLoader
loader = FileSystemLoader("templates")
environment = Environment(loader=loader)
```

2. With the environment defined, we can obtain our template:

```
tpl = environment.get_template("acl.conf.tpl")
```

3. Now that we have our template, we need to define two lists that hold the IPs we want to deny and those we want to allow. Additionally, we are going to specify the interface that we want this ACL to be configured on:

```
allowed = [
        "10.10.0.10",
        "10.10.0.11",
        "10.10.0.12"
]
```

```
disallowed = [
        "10.10.0.50",
        "10.10.0.62"
]
intf = "ethernet0"
```

4. Lastly, we need to render our template and print the rendered template back to the user:

```
out = tpl.render(allowed=allowed, disallowed=disallowed,
intf=intf)
print(out)
```

5. With the Python part done, let's open the `acl.conf.tpl` file. This is the template that is being rendered. Here, we are going to first specify the interface for our ACL and then loop over the two lists we have provided as an argument to create the necessary commands to configure our interface to allow or deny the host from the network:

```
interface {{ intf }}
ip access-group 1 in
{% for host in disallowed -%}
access-list 1 deny host {{ host }}
{%- endfor %}
{% for host in allowed -%}
access-list 1 permit host {{ host }}
{%- endfor %}
```

6. To run this script, go to your terminal and execute it with the following:

```
python3 jinja_loops.py
```

How it works...

With our Jinja2 components such as the environment and filesystem loader loaded and initialized, and the information necessary to render the template specified in our Python script, we can go ahead and render the template. This recipe does not have a lot of changes in the Python file but rather in the template.

In the template, we are using a curly bracket followed by a percentage sign to indicate that we want to use one of Jinja2's built-in logic blocks. Note that variables are indicated by using two curly braces while logic is always indicated by a single curly brace followed by a percentage sign.

Jinja2's `for` loop follows a similar syntax to the loops in Python. But while Python uses indentation to indicate what the loop body is, Jinja2 uses the `endfor` directive to indicate what block should be repeated for each entry in this list.

In the loop definition, we also added a `-` sign in front of some of the `%` signs. By default, the template engine will add whitespace. Using the `-` signs tells Jinja2 to remove this whitespace and thus makes the final configuration prettier format-wise.

There's more...

You can also add comments to your template. A comment block is indicated by a curly bracket followed by a hashtag.

Areas marked as a comment will be ignored by the template rendering system. This can be useful when working on a template where you want to comment out a part you don't need right now:

```
{# This is a comment that won't be rendered #}
```

Creating a port configuration template using if-clauses in Jinja2

With `for`-loops in our templates, we have the ability to, given a list of variables, print a modified version of the same configuration block without having to copy-paste the content.

But what if, depending on the value of the variable, we want to specify different commands? We could, as in the preceding example, use multiple lists but we can also use `if`-clauses within our template to specify any additional commands necessary in our configuration.

In this example, we are going to create a port configuration. We will specify both access and trunk ports and we will add additional commands to our trunk ports by changing the native VLAN for 802.1Q tagging.

All our ports will be described by dictionaries in our Python script.

Getting ready

Open your code editor and start by creating a file called `jinja_if.py`. Next, navigate your terminal to the same directory that you just created the `jinja_if.py` file in.

Next, in the same directory as your Python file, create a directory called `templates` if it does not already exist from a previous recipe. Inside of this directory, create a file called `ports.conf.tpl`.

How to do it...

Let's start by writing our Python script to load and render our template. In our script, we will then specify three different ports as dictionaries and hand these dictionaries over to our template.

In the template, we will then use Jinja2's `for` loops and `if` clauses to piece together our configuration:

1. Set up the Jinja2 environment. For a detailed explanation of the steps, please refer to the *Loading Jinja2 templates in Python* recipe in this chapter:

```
from Jinja2 import Environment, FileSystemLoader
loader = FileSystemLoader("templates")
environment = Environment(loader=loader)
```

2. With the environment defined, we can obtain our template:

```
tpl = environment.get_template("ports.conf.tpl")
```

3. Next, let's specify two ports. Each port will be represented by a dictionary specifying the port information, such as `type`, `slot`, port number, and interface type (`access` or `trunk`):

```
ports = []
port_1 = {
      "type": "ethernet",
      "slot": 1,
      "port_num": 1,
      "intf_type": "access"
}
ports.append(port_1)
```

```
port_2 = {
      "type": "ethernet",
      "slot": 1,
      "port_num": 1,
      "intf_type": "trunk"
}
ports.append(port_2)
```

4. Lastly, we need to render our template and print the rendered template back to the user:

```
out = tpl.render(ports=ports)
print(out)
```

5. With the Python part done, let's open the `ports.conf.tpl` file. This is the template that is being rendered. We will start by looping over each port in our ports list and specify the common information for each port. Then, we check whether the interface type of our port is trunk and if that is the case, we add an additional command that sets the native VLAN for this port to `5`:

```
{% for p in ports %}
interface {{ p['type'] }} {{ p['slot'] }}/{{ p['port_
num'] }}
switchport mode {{ p['intf_type'] }}
{% if p['intf_type'] == "trunk" %}
switchport trunk native vlan 5
{% endif %}
{% endfor %}
```

6. To run this script, go to your terminal and execute it with the following:

```
python3 jinja_if.py
```

How it works...

With our Jinja2 components such as the environment and filesystem loader loaded and initialized and the information on the ports specified in our Python script, we can go ahead and render the template. This recipe does not have a lot of changes in the Python file but rather in the template.

In the body of our `for` loop, we first print out the common configuration commands that are shared by both the trunk and access port type ports. Using the same notation we would use to retrieve the value for a key from a dictionary in Python, we can also retrieve this information in the template.

Using these dictionaries passed in from the script, we can thus render the interface information and specify the type of the port.

Next, we can then use an `if`-clause to check whether the type of our interface is trunk. Again, Jinja2 is using a similar syntax to Python here. The end of our `if`-clause is indicated similar to how the end of a `for` loop is indicated, by the `endif` directive.

There's more...

Similar to Python, Jinja2's `if`-clauses support alternative conditions such as `else if` and `else` blocks in Python:

```
{% if p['intf_type'] == "trunk" %}
      {# Add commands specific to the trunk port #}
{% elif p['intf_type'] == "access" %}
      {# Add commands specific to the access port #}
{% else %}
      {# Add commands specific to all other types #}
{% endif %}
```

Creating modular templates using Jinja2's import methods

So far, our configuration templates have all been one single file that included all the information. While this is perfectly fine for smaller configurations, it can become hard to maintain a configuration template that is thousands of lines long and spans different aspects of your configuration.

In this recipe, you'll see how to split your configuration template up into multiple sub-templates and include them in one big configuration. This technique allows you to piece your configuration together from building blocks instead of rewriting these blocks (like the configuration of an interface) every time you are creating a new configuration template.

Getting ready

Open your code editor and start by creating a file called `jinja_modular.py`. Next, navigate your terminal to the same directory that you just created the `jinja_modular.py` file in.

Next, in the same directory as your Python file, create a directory called `templates` if it does not already exist from a previous recipe. Inside of this directory, create a file called `modular.conf.tpl`. We will also need a file called `port_conf.sub.conf.tpl`.

How to do it...

Let's start by writing our Python script to load and render our template. In our script, we will then specify three different ports as dictionaries and hand these dictionaries over to our template.

In the template, we will then use Jinja2's `for` loops and `if` clauses to piece together our configuration:

1. Set up the Jinja2 environment. For a detailed explanation of the steps, please refer to the *Loading Jinja2 templates in Python* recipe in this chapter:

```
from Jinja2 import Environment, FileSystemLoader
loader = FileSystemLoader("templates")
environment = Environment(loader=loader)
```

2. With the environment defined, we can obtain our template:

```
tpl = environment.get_template("modular.conf.tpl")
```

3. Lastly, we need to render our template and print the rendered template back to the user:

```
out = tpl.render()
print(out)
```

4. With the work in our Python script done, we can now write the two templates. Our main template, `modular.conf.tpl`, will just include our sub template `port_conf.sub.conf.tpl`:

```
{% include 'port_conf.sub.conf.tpl' %}
```

5. We will put all the information regarding our port configuration into `port_conf.sub.conf.tpl` so that it looks like this:

```
interface ethernet 0/10
switchport mode access
```

6. To run this script, go to your terminal and execute it with the following:

```
python3 jinja_modular.py
```

How it works...

With our Jinja2 components such as the environment and filesystem loader loaded and initialized and the information on the ports specified in our Python script, we can go ahead and render the template. This recipe does not have a lot of changes in the Python file but rather in the template.

Instead of putting all our information into one big file, we split our configuration up into multiple smaller chunks that become more manageable. To do this, we use the `include` directive of Jinja2's template language.

There's more...

We can also use sub-templates with an `include` directive and pass variables to it. Let's modify the example from the previous recipe such that the configuration information itself is in a subtemplate and the main template only loops over all ports and then includes the sub-templates, which, in turn, is responsible for checking the variables and generating the correct configuration for the specific interface type.

Our Python script thus looks like this:

```
from Jinja2 import Environment, FileSystemLoader
loader = FileSystemLoader("templates")
environment = Environment(loader=loader)
tpl = environment.get_template("modular_ports.conf.tpl")
# Our port configuration ports = []
port_1 = {
        "type": "ethernet",
        "slot": 1,
        "port_num": 1,
        "intf_type": "access"
}
```

```
ports.append(port_1)
port_2 = {
        "type": "ethernet",
        "slot": 1,
        "port_num": 1,
        "intf_type": "trunk"
}
ports.append(port_2)
out = tpl.render(ports=ports)
print(out)
```

`modular_ports.conf.tpl` then looks like this:

```
{% for port in ports %}
{% with p=port %}
{% include "ports_config.sub.conf.tpl" %}
{% endwith %}
{% endfor %}
```

And finally, `ports_config.sub.conf.tpl` looks like this:

```
interface {{ p['type'] }} {{ p['slot'] }}/{{ p['port_num'] }}
switchport mode {{ p['intf_type'] }}
{% if p['intf_type'] == "trunk" %}
switchport trunk native vlan 5
{% endif %}
```

Using Python functions from within your template with Jinja2 filters

With `if`-clauses, loops, and the ability to pass variables into our template, we can now handle a lot of logic from right within our template file. But while the logic of Jinja2 can be quite expressive, there are limits to what you can achieve with it. Sometimes it might be easier to just use a Python function instead.

Jinja2 allows you to call custom functions on their variables to change the output of that variable. This concept is called filters and the template language itself already comes with some filters such as `length` that can be used to determine the length of a list that was passed into the template.

In this recipe, we are going to see how you can write your own Python function that can be called as a filter. In this example, we are going to write a simple function that takes an IPv4 address and converts this address in to its IPv6 equivalent using Python's built-in `ipaddress` module.

Getting ready

Open your code editor and start by creating a file called `jinja_filter.py`. Next, navigate your terminal to the same directory that you just created the `jinja_filter.py` file in.

Next, in the same directory as your Python file, create a directory called `templates` if it does not already exist from a previous recipe. Inside of this directory, create a file called `filter.conf.tpl`.

How to do it...

Let's start by writing our Python script to load and render our template. In our script, we will then create a new Python function that will be our filter. We will then inject this new filter into our Jinja2 environment and finally call it on a list of IPv4 addresses from within our template:

1. Import the required classes from the Jinja2 library as well as the `ipaddress` module, which we will use to do our conversion calculations:

```
import ipaddress
from Jinja2 import Environment, FileSystemLoader
```

2. We need to tell Jinja2 where to load our templates from. In this chapter, we will always load from the filesystem:

```
loader = FileSystemLoader("templates")
```

3. Next, we can create our Jinja2 environment. This environment is used to obtain a template object that we can later render out:

```
environment = Environment(loader=loader)
```

4. We will now need to create a Python function that will be our filter. We are going to call this function `to_ipv6()` and it will take a single argument, which is the IPv4 address. In the function, we'll use the `ipaddress` module to convert from the provided IPv4 address to an IPv6 address:

```
def to_ipv6(addr):
        raw_ipv4 = "2002::" + addr
        ipv6 = ipaddress.IPv6Address(raw_ipv4)

        return str(ipv6)
```

5. With our function defined, we need to inject it into the Jinja2 environment:

```
environment.filters["toIPv6"] = to_ipv6
```

6. With the environment defined and the filter added, we can obtain our template:

```
tpl = environment.get_template("filter.conf.tpl")
```

7. Lastly, we need to specify some IPv4 addresses that we want to convert and pass them to the `render` method:

```
addresses = ["10.10.2.10", "10.10.2.40"]
out = tpl.render(addresses=addresses)
print(out)
```

8. With the work in our Python script done, we can now write the template. In it, we are looping over all addresses and then we print both the original IPv4 address as well as the converted IPv6 address:

```
{% for addr in addresses %}
{{ addr }} -> {{ addr|toIPv6 }}
{% endfor %}
```

9. To run this script, go to your terminal and execute it with the following:

```
python3 jinja_filter.py
```

How it works...

With our Jinja2 components such as the environment and filesystem loader loaded and initialized and the information on the addresses specified, we can then modify our Jinja2 environment.

Filters in Jinja2 are simply functions that are called on the variable upon rendering the template. A filter always needs a unique name when being passed into the environment. In our case, the unique name is `toIPv6`.

In our template, we can then call this filter on a variable by using the pipe symbol. This will pass the value of the variable as the first argument to the function and the variable placeholder is replaced with whatever the filter function returns.

There's more...

You can *chain* multiple filters together. Let's assume you had two different filters, `toIPv4` and `toIPv6`. In a template, you could convert a variable first to IPv4 (using the `toIPv4` filter) and then convert that IPv4 address into an IPv6 address (using the `toIPv6` filter):

```
{{ raw_addr|toIPv4|toIPv6 }}
```

Your filter function can also receive one additional argument. Modify your code from the recipe so that the filter function now looks like this:

```
def to_ipv6(addr, prefix):
        raw_ipv4 = str(prefix) + str(addr)
        ipv6 = ipaddress.IPv6Address(raw_ipv4)

        return str(ipv6)
```

With the filter function modified, we can now provide this additional prefix information in the template:

```
{% for addr in addresses %}
{{ addr }} -> {{ addr|toIPv6("2012::") }}
{% endfor %}
```

Structuring your configuration template with blocks and template inheritance

In order to achieve modularity for our templates, we have so far relied upon including sub-templates in one main template. And while this allows us to not write the same parts of the configuration multiple times, it is also less flexible while allowing the person writing the main template to define what part to include where.

So instead of *including* sub-templates, we could also provide a base template. In this base template, we will specify *blocks* of information. We could, for example, have a block for the message of the day and another block for the port configuration.

Our actual template will then inherit this base template and just overwrite those blocks that it wants to change while keeping the defaults that are specified in the base template for those blocks that it does not overwrite.

Using this approach, we can get even more cohesion within our configuration since, no matter who writes the configuration template and in what order the blocks are overwritten, the resulting rendered configuration will always follow the format and include any additional information specified in the base template.

Getting ready

Open your code editor and start by creating a file called `jinja_base.py`. Next, navigate your terminal to the same directory that you just created the `jinja_base.py` file in.

Next, in the same directory as your Python file, create a directory called `templates` if it does not already exist from a previous recipe. Inside of this directory, create a file called `base.conf.tpl`. Additionally, we will need a file called `child.conf.tpl`.

How to do it...

Let's start by writing our Python script to load and render our template.

This template will then inherit from our base template and overwrite those blocks that it wants to change, providing modularity without having to include any sub-templates:

1. Set up the Jinja2 environment. For a detailed explanation of the steps, please refer to the *Loading Jinja2 templates in Python* recipe in this chapter:

```
from Jinja2 import Environment, FileSystemLoader
loader = FileSystemLoader("templates")
environment = Environment(loader=loader)
```

2. With the environment defined, we can obtain our template:

```
tpl = environment.get_template("child.conf.tpl")
```

3. Lastly, we need to render our template and print the rendered template back to the user:

```
out = tpl.render()
print(out)
```

4. With the work in our Python script done, we can start to specify the base template. In our `base.conf.tpl` file, we can specify the blocks like so:

```
banner motd #
{% block motd %}
Welcome to this device. This is the default message of
the day that can be changed in the configuration.
{% endblock %}
#
hostname production-{% block name_suffix %}default{%
endblock %}
```

5. Next, in `child.conf.tpl`, we need to inherit from this template by extending it:

```
{% extends 'base.conf.tpl' %}
{% block motd %}
This is a non-default message of the day!
{% endblock %}
```

6. To run this script, go to your terminal and execute it with the following:

```
python3 jinja_base.py
```

How it works...

With our Jinja2 components such as the environment and filesystem loader loaded and initialized, we are ready to render our child template based upon our base template.

In the base template, we start by specifying the blocks that we want this base template to support. The first line, `banner motd #`, will always be included, regardless of what our child templates do. After this required directive, to change the message of the day, we start by specifying our first block. A block always needs a unique name, `motd` in this case, and is defined by the `block` and `endblock` directives.

These directives mark the text that is included in this specific block.

As you can see with the name suffix block, these can also be on one line. In this case, to allow our child template to overwrite its hostname based on a specified format, blocks don't need to provide a default value, as was the case for our two preceding examples; they can also be empty.

In our child template, we start off by telling Jinja2 which template to extend. Since the base template is in the same directory as our child template, it is loaded by the same environment and it is thus enough to just specify the filename of the base template.

Be aware that you can only extend one base template per child template. Jinja2 does not support multi-inheritance.

Within the child template, we can now start overwriting the default values that are provided by specifying the blocks that we want to overwrite. Upon rendering the child template, those blocks that we have overwritten in the child template will have these new values.

All other blocks, such as the `name_suffix` block that we did not touch in our child template, will remain with the default values that were defined in our base template.

There's more...

In the preceding example, we always replaced the content of the block that was specified in our base template with new text specified in our child template.

Instead of doing that, we can also append to the text that was already specified in the base blocks by using the `super` method. When used as a template variable, this function will print the content of the block it is called in, which was specified in the base template.

Let's modify our `motd` to, instead of replacing the text specified in `base.conf.tpl`, add one more line to it.

Our `base.conf.tpl` file should then look like this:

```
banner motd #
{% block motd %}
Welcome to this device. This is the default message of the day
that can be changed in the configuration.
{% endblock %}
#
```

Our `child.conf.tpl` file now needs to, when overwriting the `motd` block, call the `super` method to still keep the configured message while adding to it.

`child.conf.tpl` thus should look like this:

```
{% extends 'base.conf.tpl' %}
{% block motd %}
{{ super() }}

This line is then added by the child template.
{% endblock %}
```

As you can see, we call the `super()` method as the first line, after specifying which block we want to modify. This way, the *this line is added by the child template* text is appended after the original message.

You could also call the `super` method *after* the additional text you specified.

4

Configuring Network Devices Using Netmiko

In *Chapter 2, Connecting to Network Devices via SSH Using Paramiko*, we saw how to connect to a network device via SSH using the Paramiko library. While we can do pretty powerful things such as executing commands against our devices using Paramiko, it was built as an SSH library, not a library to connect to and configure network devices. Enter **netmiko**, a library purpose-built to connect to and configure network devices using Python.

The netmiko library is build on top of Paramiko, hence the similarity in name, which abstracts away some of the quirks of dealing with SSH connections, including differing escape codes, and is purpose-built for connecting to and configuring network devices using either an SSH or telnet connection. This purpose-built abstraction becomes particularly useful when dealing with devices from different vendors and not with a homogeneous single-vendor deployment. Additionally, netmiko provides a nice set of helper functions that allow you to carry out common workflows, such as connecting to a device and deploying a configuration from a file, with fewer lines of code.

In this chapter, we are going to learn how to use the netmiko library to connect to our devices, issue commands, retrieve output and copy files all from within Python. With the techniques learned in this chapter, you'll be able to automate existing workflows that involve applying one or more commands to a network device from Python as well as retrieve information from your network devices in a structured way.

The recipes in this chapter will cover the following areas:

- Connecting to a network device using netmiko
- Sending commands using netmiko
- Retrieving command outputs as structured Python data using netmiko and Genie
- Gathering facts using netmiko
- Connecting to multiple devices
- Creating and applying a configuration template with Jinja2 and netmiko
- Copying files to a device using netmiko
- Escalating privileges with netmiko
- Authenticating using public-private keys with netmiko
- Handling commands that prompt for information using netmiko

Technical requirements

For this section and the remainder of the book, you'll need an installation of Python. Specifically, you'll need a Python interpreter of version 3.6.1 or higher. This book makes use of the language constructs of Python 3 and thus is incompatible with Python 2.x. Please also install the netmiko package (`python3 -m pip install netmiko`). We are using the latest version of netmiko at the time of writing, version 3.3.3. You may install this exact version by issuing `python3 -m pip install netmiko==3.3.3`. All code examples have been developed and tested on a Mac running macOS version 10.15.4.

You also want a code editor. Popular choices include Microsoft Visual Studio Code or Notepad++. Additionally, you'll need a device (virtual or physical) that you can log in to via SSH.

You can view this chapter's code in action here: `https://bit.ly/3yHL5Qi`

Connecting to a network device using netmiko

When dealing with a network device and its connection in netmiko we will generally deal with `ConnectHandler`. This is our central interface to open a connection to a device and write commands to the device as well as reading the output back.

In this recipe, you will see how to programmatically open an SSH connection by creating an instance of netmiko's `ConnectHandler`.

Getting ready

Open your code editor and start by creating a file called `connect.py`. Next, navigate your terminal to the same directory that you just created the `connect.py` file in.

How to do it...

Let's start by importing the required classes from the netmiko library. We will then set up a dictionary that contains our connection details and then initiate a connection to the device we just specified.

Follow these steps to establish a connection from your Python script to a network device using netmiko:

1. Import `ConnectHandler` from `netmiko`:

```
from netmiko import ConnectHandler
```

2. Create a dictionary that will contain our connection details. In this example, we are connecting to a Cisco IOS device. See the *How it works* section for a list of device types from other vendors. Make sure to use the exact same dictionary keys (such as `device_type` or `host`) when specifying your connection information:

```
connection_info = {
        'device_type': 'cisco_ios',
        'host': '<insert your host here>',
        'port': <insert your port number here>,
        'username': '<insert your username here>',
        'password': '<insert your password here>'
}
```

3. Next, we are going to use a Python context manager to open a connection, which is similar to opening a file:

```
with ConnectHandler(**connection_info) as conn:
        print("Successfully connected!")
```

4. To run this script, go to your terminal and execute it with the following command. The output should be the same as the following example:

```
python3 connect.py
Successfully connected!
```

How it works...

In this example, we first loaded `ConnectHandler` from netmiko into our script. This is the central class we are going to use when interacting with our network device from a netmiko perspective.

Next, we specified the connection details in a dictionary. The first entry, `device_type`, is a hint to the netmiko library about what kind of device we are dealing with.

Based on this hint, netmiko will handle the SSH connection differently. For example, different vendors may have different escape sequences or handle prompts differently and netmiko accounts for these differences in the background so that you as a developer don't have to worry about it.

While writing this book in January 2021, the netmiko library supports 105 different device types. A complete list can be found in this source code file: `https://github.com/ktbyers/netmiko/blob/master/netmiko/ssh_dispatcher.py#L104`.

While the entire list of 105 handlers would be excessive to reiterate here, the following are the handlers for the most common (network) device vendors:

- `cisco_ios`, `cisco_xe`, and `cisco_xr` for connecting to Cisco IOS devices
- `cisco_nxos` for connecting to devices running the Cisco NX-OS operating system
- `linux` for connecting to generic Linux devices such as servers
- `juniper`, `juniper_junos`, and `juniper_screenos` for connecting to devices from Juniper
- `arista_eos` for connecting to Arista devices
- `hp_comware` and `hp_procurve` for connecting to devices from HP
- `huawei`, `huawei_smartax`, `huawei_olt`, and `huawei_vrpv8` for connecting to devices from Huawei

If the vendor that produced your device is not in the preceding bullet list, please refer to the link mentioned prior to the preceding list to check the entire listing of available device types.

With the device type defined, we can now go ahead and specify the familiar connection attributes, such as `host`, `username`, `password`, and `port`.

Next, we are using the Python construct of a context manager. A context manager is a Python construct that allows us to carry out two related operations one after another without having the developer write excessive boilerplate code. Essentially, context managers allow library developers to specify code to be executed when the object is created, before the user-provided code is executed, and after the user-provided code is executed.

The most common example of where context managers can lead to fewer lines of code is when opening and closing a file. When opening a file, we generally want to do something with said file, such as reading or writing to or from it, and then close the file. The context manager in Python allows developers of a library to offer this exact functionality to the users using the `with` keyword. In the previous chapter, we saw files being opened with this method as well:

```
with open("file.txt", "r") as f:
        lines = f.readlines()
        for line in lines:
                print(line)
```

This code is equivalent to the following code snippet:

```
f = open("file.txt", "r")
lines = f.readlines()
for line in lines:
        print(line)
f.close()
```

netmiko also defines a context manager for its `ConnectHandler` so we can use the `with` statement to use it. With this, netmiko also takes care of closing the connection to the device once we have executed all our commands.

In the previously discussed example, we are using the double asterisk operator to unpack the information specified in the dictionary and pass that information on as keyword arguments to `ConnectHandler`. Therefore, it is important to use the exact same spelling since it refers to the `connection_info` dictionary. Please refer to the *There's more* section of this recipe if you want to learn more about the syntax used here to pass the arguments from a dictionary.

Instead of the asterisk syntax we could have also written the following:

```
connection_info = {
        'device_type': 'cisco_ios',
        'host': '<insert your host here>',
        'port': <insert your port number here>,
        'username': '<insert your username here>',
        'password': '<insert your password here>'
}
with ConnectHandler(device_type='cisco_ios',
                    host='<your_host>',
                    port=<insert your port number here>,
                    username='<insert your username here>',
                    password='<insert your password here>') as conn:
        print("Successful connection!")
```

As you can see, this makes the example significantly longer and harder to read. Also, by storing the connection information in the dictionary, you could do things such as storing the access information for all your devices in JSON files and then read them back as dictionaries when needing to connect to a certain device.

There's more...

The `**`syntax seen in the preceding section is often used in conjunction with a concept called **variadic** arguments. Variadic arguments allow a function to receive an arbitrary number of arguments from the user without specifying the arguments.

Think for example of a function that should be able to receive an arbitrary number of integers as an argument and then sum them. You could come up with the idea to specify a function with the same name, but different parameters. So, you could imagine writing something like this:

```
def sum_up(a):
        return a
def sum_up(a, b):
        return a + b
```

This won't work because Python does not support **function overloading** – the process of having the same function name with different argument sets. Python does not support function overloading officially. If you are interested in hacking Python to support this have a look at the concept of virtual namespaces. Together with decorators, you can build this functionality into Python. This is beyond the scope of this book, however. With variadic arguments we can build this functionality quite elegantly.

Please refer to the following code to understand how variadic arguments work:

```
def sum_up(*nums):
    print(type(nums))
    result = 0
    for num in nums:
        num += num
    return num
sum_up(1, 2, 3, 4)
```

As you can see, the type of the `nums` variable is a tuple. You can think of a tuple as a list that can't be changed after it is created. So, we are essentially passed a list of all the arguments that the function has received. Mind the `*` we used to specify this to be a variadic argument.

Similarly, we can use the `**` syntax that we used in our example to unpack a dictionary to keyword arguments to specify that we want to receive a dictionary with all the keyword arguments as follows:

```
def sum_up(**kwargs):
    print(str(kwargs.keys())
sum_up(hello="world")
```

While you can name these two types of arguments however you like, the names `*args` and `**kwargs` have established themselves as the quasi-standard.

Sending commands using netmiko

Now that we have successfully established a connection to our network device using netmiko and `ConnectHandler`, we want to be able to send commands to our device. In Paramiko, the process of sending and receiving the output was rather cumbersome. Luckily, netmiko provides an easy-to-use abstraction over Paramiko that lets us issue a command and read back the output with just one line of code.

In this recipe, you will see how to programmatically open an SSH connection by creating an instance of netmiko's `ConnectHandler`, send a command to the remote device, and retrieve the output of the command as a string.

Getting ready

Open your code editor and start by creating a file called `send_command.py`. Next, navigate your terminal to the same directory that you just created the `send_command.py` file in.

How to do it...

Let's start by importing the required classes from the netmiko library. We will then set up a dictionary that contains our connection details and then initiate a connection to the device we just specified. With the connection covered we can then proceed and issue our command and print the output back to the user.

Follow these steps to connect to a network device and send commands using netmiko:

1. Import `ConnectHandler` from netmiko:

```
from netmiko import ConnectHandler
```

2. Create a dictionary that will contain our connection details. In this example, we are connecting to a Cisco IOS device. See the *How it works* section in the *Connect to a network device using netmiko* recipe for a list of device types from other vendors. Make sure to use the exact same dictionary keys (such as `device_type` or `host`) when specifying your connection information:

```
connection_info = {
        'device_type': 'cisco_ios',
        'host': '<insert your host here>',
        'port': <insert your port number here>,
        'username': '<insert your username here>',
        'password': '<insert your password here>'
}
```

3. Next, we are going to use a Python context manager to open a connection similar to how we can open a file. Inside of that context manager, we'll then use the `send_command()` method of `ConnectHandler` to send our `show interfaces` command to the device and print back the output:

```
with ConnectHandler(**connection_info) as conn:
        out = conn.send_command("show interfaces")
        print(out)
```

4. To run this script, go to your terminal and execute it with the following command. The output should be the same as if you had connected to the device by SSH directly and issued the following command:

```
python3 send_command.py
```

How it works...

We first import the `ConnectHandler` class from netmiko. Next, we specify the connection details for our device, the host, port, username, and password, as well as the device type. Netmiko uses the device type internally to provide a common interface to send commands to devices and operating systems from different vendors. Please see the *How it works* section in the *Connect to a network device using netmiko* recipe for a list of supported device types.

With our context manager established, we can now send the command – in this example, a `show interface` command on a Cisco IOS device using the `send_command()` method.

The output returned from this method is a string that contains the output from the command you issued. We then print out that command for the user and the context manager takes care of closing the connection.

Retrieving command outputs as structured Python data using netmiko and Genie

In the previous recipe (*Sending commands using netmiko*), we saw how to send a command and retrieve the output of said command as a string.

While the text output might be perfectly suitable for a human to look at, and understand the different parts, computers have a hard time understanding plaintext.

What if, for example, we want to build a script that prints out all the different interfaces that are available on our networking device? With plaintext, this is going to involve a lot of text manipulation. Since turning the output of CLI commands into structured data is a cornerstone of network automation, libraries have been developed to address this exact issue.

One such library is **Genie**. Genie (together with pyATS) is an open source project originally developed and currently maintained by Cisco. While it has its roots as an internal Cisco library, it supports the devices from other vendors through its open interfaces. Genie and pyATS have an active community with new parsers constantly being added into the new releases.

We will see a lot more of the features that Genie and pyATS provide in *Chapter 7, Automate Your Network Tests and Deployments with pyATS and Genie*, but for now we'll be content with Genie being a very powerful parser that can convert the textual output of a command into structured Python data.

Getting ready

Open your code editor and start by creating a file called `get_interfaces.py`. Next, navigate in your terminal to the same directory that you just created the `get_interfaces.py` file in.

When installing netmiko via pip, it does not come preinstalled with pyATS and Genie. So we'll need to install these two additional packages by issuing `python3 -m pip install pyats genie`. The recipes in this book were developed against version 21.1 of pyATS and version 21.1.3 of Genie. If you wish to install the same versions, you can use the following commands: `python3 -m pip install pyats==21.1` and `python3 -m pip install genie==21.1.13`. Please note that Genie and pyATS are not supported on Windows. If you use Windows as your main operating system you may need to either install Linux in a virtual machine or use the Windows Subsystem for Linux 2 to use pyATS/Genie.

How to do it...

After installing the required new packages, we can go ahead and connect to our device, issue the `show interfaces` command and then use Genie to parse the textual output into a Python dictionary. With that Python dictionary in hand, we'll first print out the entire content of the dictionary to see how Genie structures the output and then print out just the list of interfaces.

Using the following steps you can retrieve the output of a command as structured Python data instead of plaintext:

1. Import the `ConnectHandler` from netmiko. Additionally, we'll import the pretty print module that allows us to print a dictionary with better formatting:

```
from netmiko import ConnectHandler
import pprint
```

2. Create a dictionary that will contain our connection details. In this example, we are connecting to a Cisco IOS device. See the *How it works* section in the *Connect to a network device using netmiko* recipe for a list of device types from other vendors. Make sure to use the exact same dictionary keys (such as `device_type` or `host`) when specifying your connection information:

```
connection_info = {
        'device_type': 'cisco_ios',
        'host': '<insert your host here>',
        'port': <insert your port number here>,
        'username': '<insert your username here>',
        'password': '<insert your password here>'
}
```

3. Next, we are going to use a Python context manager to open a connection, similar to how we open a file. Inside of that context manager, we'll then use the `send_command()` method of the `ConnectHandler` to send our `show interfaces` command to the device and print back the output. Notice how we have set the `use_genie` flag of `send_command` to `True`:

```
with ConnectHandler(**connection_info) as conn:
        out = conn.send_command("show interfaces",   use_
genie=True)
```

4. Next, we'll print out the entire dictionary we received using prettyprint. Make sure that you are still within the context manager (one indention level):

```
        pprint.pprint(out)
```

5. And finally, we iterate over all the items within the dictionary. The keys of this specific dictionary will be the names of the interfaces. The output of course depends on the specific command you are issuing:

```
        for interface in out.keys():
                print(interface)
```

6. To run this script, go to your terminal and execute it with the following command:

```
python3 get_interfaces.py
```

This command will give the following output:

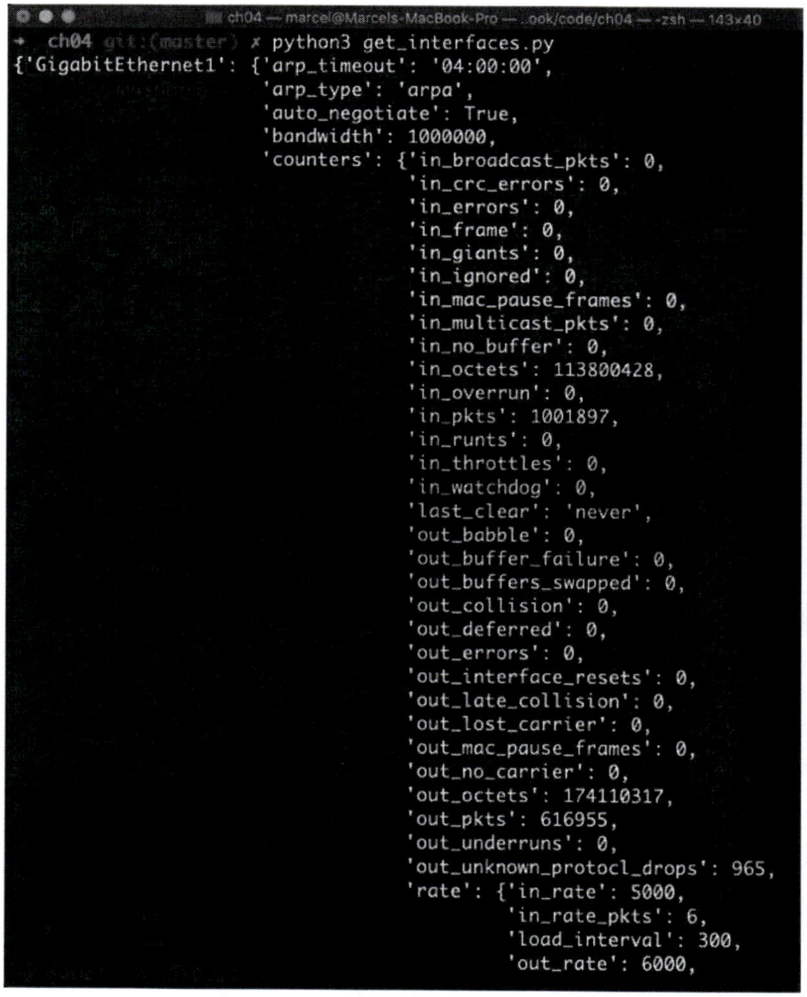

Figure 4.1 – Excerpt of the returned output

The preceding screenshot shows an excerpt of the output from the returned dictionary printed by the `prettyprint` library. As you can see, the dictionary is nicely formatted and properly indented, making it easier to read through.

How it works...

With the netmiko connection established (please have a look at the *How it works* section of the *Send commands using netmiko* recipe for a detailed explanation) we can send our command again and retrieve the output.

The only difference to the way we invoked `send_command()` in the *Send commands using netmiko* recipe was that we toggled the `use_genie` flag to `True`. Internally, this invoked the Genie parser, and the returned object is no longer the plaintext returned by the network device itself but the parsed output in the form of a dictionary!

We can then use the built-in pretty print library to get a formatted version of the dictionary printed back to us. With our `show interfaces` command the Genie parser put the names of the interfaces as the key to a nested dictionary. Contained in that dictionary are all the details returned by the command in structured Python data. You could for example find the bandwidth of an interface called GigabitEthernet1 using `out['GigabitEthernet1']['bandwidth']`. Notice how Genie has also converted numbers to numbers and Booleans to Python Booleans to make it easy for you to do comparisons or check if a certain setting is set to `true` or `false`.

With our knowledge of how the data returned in the dictionary is structured, we can go ahead and print our desired list of interfaces by looping over all the keys in the dictionary returned by Genie.

Genie is constantly being updated and new parsers added. You may find an up-to-date list of all parsers currently included at `https://pubhub.devnetcloud.com/media/genie-feature-browser/docs/#/parsers`.

In the explorer on Genies documentation page, in the top right you can specify the operating system you are interested in and then either scroll through the list or use the search bar to verify that the commands you are using are supported by Genie.

Gathering facts using netmiko

In the previous recipe (*Retrieving command outputs as structured Python data using netmiko and Genie*), we saw how to retrieve the output of a command as structured data in Python. In this recipe, we will build upon this functionality to get a profile of all our interfaces. This will show you how to use the parsed data to quickly print out and identify the relevant information.

For this example, we'll be using the `show interfaces` command again and, for each interface, we will retrieve and print out the following:

- The interface name
- The status of the interface (enabled or not enabled)
- The physical address of the interface
- The duplex mode
- Any counters that are above 0

Getting ready

Open your code editor and start by creating a file called `get_facts.py`. Next, navigate your terminal to the same directory that you just created the `get_facts.py` file in.

When installing netmiko via pip, it does not come preinstalled with pyATS and Genie. So we'll need to install these two additional packages by issuing `python3 -m pip install pyats genie`. The recipes in this book were developed against version 21.1 of pyATS and version 21.1.3 of Genie. If you wish to install the same versions, you can use the following commands: `python3 -m pip install pyats==21.1` and `python3 -m pip install genie==21.1.13`.

How to do it...

Using the following steps you can retrieve the output of our `show interfaces` command as structured Python data that will then be parsed further:

1. Import `ConnectHandler` from netmiko. Additionally, we'll import the pretty print module that allows us to print a dictionary with better formatting:

```
from netmiko import ConnectHandler
import pprint
```

2. Create a dictionary that will contain our connection details. In this example, we are connecting to a Cisco IOS device. See the *How it works* section in the *Connect to a network device using netmiko* recipe for a list of device types from other vendors. Make sure to use the exact same dictionary keys (such as `device_type` or `host`) when specifying your connection information:

```
connection_info = {
        'device_type': 'cisco_ios',
        'host': '<insert your host here>',
```

```
        'port': <insert your port number here>,
        'username': '<insert your username here>',
        'password': '<insert your password here>'
  }
```

3. Next, we are going to use a Python context manager to open a connection, similar to opening a file. Inside of that context manager, we'll then use the `send_command()` method of the `ConnectHandler` to send our `show interfaces` command to the device and print back the output. Notice that we have set the `use_genie` flag of `send_command` to `True`:

```
with ConnectHandler(**connection_info) as conn:
        out = conn.send_command("show interfaces",   use_
genie=True)
```

4. With our structured data retrieved we can now iterate over each of the interfaces and print the information we want. Make sure that you are still within the context manager (one indention level):

```
        for name, details in out.items():
            print(f"{name}")
            print(f"- Status: {details.get('enabled',
None)}")
            print(f"- Physical address: {details.get('phys_
address', None)}")
            print(f"- Duplex mode: {details.get('duplex_
mode', None)}")
```

5. With our basic information printed out, we can now iterate over all counters and their values to check if any of those counters is over 0. Make sure that you are still within the context manager (first indention level) as well as within the `for` loop from the interfaces (second indention level):

```
            for counter, count in details.get('counters',
{}).items():
                if isinstance(count, int):
                    if count > 0:
                        print(f"- {counter}: {count}")
                elif isinstance(count, dict):
                    for sub_counter, sub_count in count.
items():
```

```
                if sub_count > 0:
                    print(f"- {counter}::{sub_
counter}: {sub_count}")
```

6. To run this script, go to your terminal and execute it with the following command:

```
python3 get_facts.py
```

How it works...

With the netmiko connection established (please have a look at the *How it works* section of the *Send commands using netmiko* recipe for a detailed explanation) we can send our command again and retrieve the output as structured data (please see the *How it works* section of the *Retrieving command outputs as structured Python data using netmiko and Genie* recipe for a detailed explanation of how this is achieved).

We then iterate over each of our interfaces and their associated details. We first print out the static information such as the interface name and status. Note how we are using for example the `details.get("phys_address", None)` function of the dictionary instead of directly accessing it using the square brackets (that is, writing `details["phys_address"]`). The benefit of using the `get` method is that we can provide a default value, in our case `None`, that is returned in case the key is not found. This is to prevent a Key Error. Key errors occur when we are trying to access the key-value pair of a dictionary where the key does not exist, especially when dealing with output from devices. It is thus advisable to use this get method. Some interfaces might have an IPv4 address configured. Others might not have that particular field specified. Using the `get` method, we can be sure that our code does not stop executing but rather just prints the default value.

With our basic information printed, we can next iterate over the available counters. Counters can come in two different formats. They are either a pair of a counter name (for example `in_crc_errors`) and the associated value (for example `0`) or they can be a counter name followed by a dictionary that contains the sub counter name and sub counter value pairs. We use the built-in `isinstance()` method to check if our counter is a number or if we are dealing with a dictionary and thus have mapping of sub counter name and sub counter value pairs.

The exact output of the structured data depends on how Genie parses the command. You can see the returned structure by having a look at the parsers available. Genie is constantly being updated and new parsers added. You can find an up-to-date list of all parsers currently included at `https://pubhub.devnetcloud.com/media/genie-feature-browser/docs/#/parsers`.

In this explorer, in the top right you can specify the operating system you are interested in and then either scroll through the list or use the search bar to verify that the commands you are using are supported by Genie and how the output is structured. Alternatively, you can always run the command with `use_genie` set to `True` and then use the pretty print library (as shown in the *Retrieving command outputs as structured Python data using netmiko and Genie* recipe) to see the structure of the actual output for yourself. In the pretty printer output you can then identify the keys and the types of their values.

There's more...

The `get()` method of a dictionary returns `None` by default, so in the preceding code we could have omitted the second argument, writing `details.get("enabled")` instead of `details.get("enabled", None)` since `None` is the default return value. If you want to return a different default return value – let's say you want to return the string `N/A` (commonly used as an abbreviation of *Not Applicable*) – you would write `details.get("enabled", "N/A")`.

Connecting to multiple devices

So far, we have always only dealt with a single device and our connection info was always specified directly in the Python script. It is quite a common use case though to issue the same set of commands or the same configuration to an entire fleet of devices.

Using a script to apply the same command automatically instead of logging into each device manually not only saves time but also prevents configuration drift since all devices are guaranteed to be issued the exact same commands.

In this recipe, we are going to specify the connection information for our devices in the form of a JSON file. Based on the connection information stored in this JSON file we'll then connect to each of these devices, issue a command (`show running-config` in this example), retrieve the output, and save it to a file.

Getting ready

Open your code editor and start by creating a file called `connect_multiple.py`. Next, navigate your terminal to the same directory that you just created the `connect_multiple.py` file in.

We'll also need a file to store our connection details. So, in the same directory as your `connect_multiple.py` file, create a file called `connections.json`.

How to do it...

We'll first import the built-in json library as well as the `netmiko` module. Next, we are going to specify our connection details in our `connections.json` file, read that file in our Python script, and open up a connection based on the provided details. We'll issue the `show running-config` command in each of the devices and save the output into a text file.

Using the following steps we can read connection details from a JSON file, connect to each specified device, execute a command, and save the output of the command per device:

1. Open the `connections.json` file in your code editor. For each device, we are going to define the following structure. Note how the nested dictionary with the key connection has the exact same keys as those we previously defined in our Python dictionary:

```
[
  {
    "name": "device-1",
    "connection": {
       "device_type": "<insert your device type here>",
        "host": "<insert your host here>",
        "port": <insert your port number here>,
        "username": "<insert your username here>",
        "password": "<insert your password here>"
    }
  },
{
    "name": "device-2",
    "connection": {
       "device_type": "<insert your device type here>",
        "host": "<insert your host here>",
        "port": <insert your port number here>,
        "username": "<insert your username here>",
        "password": "<insert your password here>"
    }
  }
]
```

2. With all your devices specified you can open the `connect_multiple.py` file. We first are going to import the required libraries:

```
import json
from netmiko import ConnectHandler
```

3. Next, we need to read the connection details from our JSON file:

```
devices = []
with open("connections.json", "r") as fh:
        devices = json.load(fh)
```

4. We can now iterate over each device in the devices list using the connection information stored in the JSON file, issue the defined command, and then save the output to a new file based on the name we gave the device in the JSON file:

```
CMD = "show running-config"

for device in devices:
            file_name = f"{device['name']}.out"
    print(f"Retrieving config for {device['name']}")
    with ConnectHandler(**device['connection']) as conn:
                out = conn.send_command(CMD)
        with open(file_name) as f:
                f.write(out)
```

5. To run this script, go to your terminal and execute it with the following command:

```
python3 connect_multiple.py
```

You should see output printed acknowledging that the command is being issued for each of your devices specified in the `connections.json` file.

How it works...

In this recipe, we are defining our connections in a JSON file instead of putting them directly into our Python script. In the JSON file, we defined a list, as indicated by the opening and closing square brackets, of dictionaries. Each of our dictionaries representing a device has a name and another nested dictionary with the key connection. Contained in this nested connection dictionary is the connection information for the device, including details such as the host, port, username, password, and device type required by netmiko to connect successfully.

With the JSON file created, we next read that information back into Python by opening the file in Python and then reading and parsing the content from JSON into a Python list of dictionaries using the built-in json module.

With our device data loaded, we can now specify the command we want to issue on all our devices and then loop over all the dictionaries representing a device contained within the devices list.

For each of our devices, we then create a filename for the output based on the name we provided in the JSON file. We then open a `ConnectHandler` with the connection details contained within the `connection` sub-dictionary and execute the command specified previously on the device.

Finally, we open a new file handler to write the contents of our output, in this case the running configuration of the device, to a file.

With this simple script, you could already create a full backup of all your configurations. This could be useful for example when doing a big update and you would want to have the current state of the configuration preserved in case something goes wrong and you need to revert to this version of your configuration.

There's more...

Depending on the size of your network it might be difficult to keep the credentials and connection details for each of your network devices in `.json` files. While programmatically consumable files like the ones shown in this recipe are preferable to keeping your connection details in a `.pdf` document, large networks might require a centralized storage for all information. This is where **IP Address Management** (**IPAM**) tools come into play. Popular solutions such as the open source IPAM Netbox (`https://netbox.readthedocs.io/en/stable/`) allow you to have a centralized storage for all your device details. Netbox then offers a REST API that can be used to retrieve the information. We will discuss REST APIs and how to consume them with Python in *Chapter 9, Consuming Controllers and High-Level Networking APIs with requests.*

Creating and applying a configuration template with Jinja2 and netmiko

With the recipes so far, we have only read back information from our network devices. What we have not done so far is apply a new configuration to our device.

In *Chapter 3, Building Configuration Templates Using Jinja2*, we saw how we can use the Jinja2 templating language to build a new configuration file based on an existing template. In this recipe, we are going to combine Jinja2's ability to render a configuration template into an actual configuration file and netmiko's ability to apply a file as a new configuration to a network device to go from a configuration template to a configuration that is applied to a real device.

Getting ready

Open your code editor and start by creating a file called `apply_config.py`. Next, navigate your terminal to the same directory that you just created the `apply_config.py` file in.

Additionally, we'll need a Jinja2 template, so in the same directory as your Python file, create a directory called `templates`. Inside of this directory, create a file called `acl.conf.tpl`.

We'll be reusing the same template we rendered to a file in *Chapter 3, Building Configuration Templates Using Jinja2*, in the *Use for loops in Jinja2 to configure an access list* recipe.

If you have not installed the Jinja2 package in *Chapter 3, Building Configuration Templates Using Jinja2*, please go ahead and do so now. You can install the newest version of Jinja2 using `python3 -m pip install jinja2`. At the time of writing, the current version is version `2.11.2`. You can install this specific version by using the following command: `python3 -m pip install jinja2==2.11.2`.

How to do it...

We'll first create our configuration template, load that template into our Jinja2 environment, and render it. After writing the rendered template to a file we'll then use netmiko to open a connection to our network device and apply the previously rendered configuration to our device as follows:

1. Open the `acl.conf.tpl` file. This is the template that is being rendered. Here, we are going to first specify the interface for our ACL and then loop over the two lists we have provided as an argument to create the necessary commands to allow or deny the host from the network:

```
interface {{ intf }}
ip access-group 1 in
{% for host in disallowed %}
```

```
access-list 1 deny host {{ host }}
{% endfor %}
{% for host in allowed %}
access-list 1 permit host {{ host }}
{% endfor %}
```

2. Open the `apply_config.py` file. We'll start by importing the required libraries (netmiko and Jinja2):

```
from netmiko import ConnectHandler
from jinja2 import Environment, FileSystemLoader
```

3. With our libraries imported we can define the connection details of our device:

```
connection_info = {
        'device_type': 'cisco_ios',
        'host': '<insert your host here>',
        'port': <insert your port number here>,
        'username': '<insert your username here>',
        'password': '<insert your password here>'
}
```

4. Next, we need to set up our Jinja2 environment to render the template:

```
loader = FileSystemLoader("templates")
environment = Environment(loader=loader)
tpl = environment.get_template("acl.conf.tpl")
```

5. Now that we have our template, we need to define two lists that hold the IPs we want to deny and those we want to allow. Additionally, we are going to specify the interface that we want this ACL to be configured on in the following manner:

```
allowed = [
        "10.10.0.10",
        "10.10.0.11",
        "10.10.0.12"
]
disallowed = [
        "10.10.0.50",
        "10.10.0.62"
```

```
]
intf = "ethernet0"
```

6. With the information needed to render our template in place, we can create a configuration from our template and write the output to a file:

```
out = tpl.render(allowed=allowed, disallowed=disallowed,
intf=intf)
with open("configuration.conf", "w") as f:
        f.write(out)
```

7. Now that our template is rendered and written to a file called `configuration.conf`, we can establish a connection using `ConnectHandler` from netmiko. Instead of using the `send_command()` method we are going to use the `send_config_from_file()` method that takes the path to a configuration file as an argument as follows:

```
with ConnectHandler(**connection_info) as conn:
        out = conn.send_config_from_file("configuration.
conf")
```

8. To run the script, go to your terminal and execute the following command:

```
python3 apply_config.py
```

How it works...

In this recipe, we first create the configuration template, set up the Jinja2 environment, and then render the template. Please refer to *Chapter 3, Building Configuration Templates Using Jinja2*, in the *Use for loops in Jinja2 to configure an access list* recipe, for a more detailed explanation as to how the setup of Jinja2 works.

With our template rendered and written into the new configuration file, we can now go ahead and open a connection to our network device. With the connection established, we can then use the convenient `send_config_from_file()` method.

This method takes the path to a configuration file as an input, reads it line by line, and applies the commands to the output device.

There's more...

In this example, we rendered the configuration from a template file and stored that template file on our hard disk before having netmiko read the file back and apply it. If you have a list of configuration commands that you would like to issue and you already have that list as a Python list, you can make this easier in the following way:

```
commands = [
    'interface ethernet0',
    ' ip access-group 1 in',
    'access-list 1 deny host 10.10.10.10'
]
with ConnectHandler(**connection_info) as conn:
     out = conn.send_config_set(commands)
```

Copying files to a device using netmiko

Transferring files to and from a network device is a common use case. Be it the transfer of a new firmware from your computer to a network device, or the download of log files from the network device back to your computer for further analysis, it is always useful to be able to sync files with a device.

In this recipe we'll see how netmiko has integrated the secure copy (or `scp`) protocol that works on top of SSH to allow us to easily transfer files to and from a device.

Getting ready

Open your code editor and start by creating a file called `transfer_files.py`. Next, navigate your terminal to the same directory that you just created the `transfer_files.py` file in.

You'll also need a file, for example a new firmware or just a text file, that you want to transfer to your device. If you don't have a file ready that you could transfer, feel free to just create an empty text file called `test_upload.txt`. You'll also have to make sure that SCP is enabled on your device. Please refer to the manual of your device to find out how to enable secure copy on your operating system.

How to do it...

In this recipe, we'll first import the required classes and methods from the `netmiko` module. With that in place we'll create our connection information that will include file system information. With this, we can use netmiko to transfer our local file to the network device.

Follow these steps to establish a connection to the specified device and transfer a file to it:

1. Import the `ConnectHandler` from netmiko as well as the `file_transfer` function:

```
from netmiko import ConnectHandler, file_transfer
```

2. Next, we need to specify the required connection information:

```
connection_info = {
        'device_type': 'cisco_ios',
        'host': '<insert your host here>',
        'port': <insert your port number here>,
        'username': '<insert your username here>',
        'password': '<insert your password here>',
}
```

3. With our connection information defined, we can open a `ConnectHandler` to our remote device and use the `file_transfer` function imported before to upload our local file:

```
with ConnectHandler(**connection_info) as conn:
    ret = file_transfer(conn,
                   source_file="test_upload.txt",
                   dest_file="test_on_device.txt",
                   file_system="<insert target filesystem
here>",
                   direction="put")
    for k, v in ret.items():
        print(f"{k}: {v}")
```

4. Run the script by executing the following command:

```
python3 transfer_file.py
```

Once the script has run you may want to verify that the file has indeed been transferred by logging into the device via SSH.

How it works...

In this recipe, we are using the `file_transfer` function provided by netmiko to upload a file from our local machine to the remote device using SCP.

First, we define our connection details for the device we want to upload our file to. Next, we create a new `ConnectHandler` within a context manager. This `ConnectHandler` is then passed as the first argument into the `file_transfer()` function.

The `file_transfer` function takes the following arguments:

- `source_file` is the path to the local file we want to upload.
- `dest_file` is the name of the file in our destination.
- `file_system` allows us to specify which file system we want to write to on the target device, such as `bootflash`, `flash`, or `disk0`.
- `direction` specifies whether we are transferring to or from a device. To transfer files from the local machine to a device, we use `put`, and to transfer a file from a remote device to our local device, we use the `get` direction.

The `file_transfer` function returns a dictionary that contains three Boolean flags related to our remote file:

- `file_exists` indicates that the file does now exist on the target device.
- `file_transfered` indicates whether the file actually had to be transferred or whether it was already present on the target device.
- `file_verified` indicates whether the **MD5 checksum** of the transferred file matches the MD5 signature of the local file.

There's more...

In the example, you saw how you can upload a file to a device. You can also do the reverse by setting the direction to `get`.

By default, the transferred file will be verified by comparing the **MD5 hash**. This verification can take a long time and thus you can disable it by setting the `disable_md5` flag to `True`.

We can have the `file_transfer` function overwrite files by setting the `overwrite_file` flag to `True`.

Escalating privileges with netmiko

When dealing with configuration operations on a network device, we might need to enter a mode that has more privileges. This enable mode can, and should, be secured by a password.

In the examples so far, we have assumed that this was not the case and that everyone can just enter **enable** mode without being prompted for a password. In this recipe, we are going to see how you can provide netmiko with the secret when connecting to the device and then enter enable mode before retrieving the running configuration.

Getting ready

Open your code editor and start by creating a file called `enter_enable.py`. Next, navigate your terminal to the same directory that you just created the `enter_enable.py` file in.

Make sure that your device is configured to require a secret when trying to enable.

How to do it...

Let's start by importing the required classes from the netmiko library. We will then set up a dictionary that contains our connection details and then initiate a connection to the device we just specified. With the connection covered we can then proceed and issue our command and print the output back to the user.

Follow these steps to programmatically enter enable mode before issuing a command:

1. Import `ConnectHandler` from netmiko:

```
from netmiko import ConnectHandler
```

2. Create a dictionary that will contain our connection details. In this example, we are connecting to a Cisco IOS device. See the *How it works* section in the *Connect to a network device using netmiko* recipe for a list of device types from other vendors. Make sure to use the exact same dictionary keys (such as `device_type` or `host`) when specifying your connection information. Notice how we have added a key-value pair for the `secret`:

```
connection_info = {
        'device_type': 'cisco_ios',
        'host': '<insert your host here>',
        'port': <insert your port number here>,
        'username': '<insert your username here>',
        'password': '<insert your password here>',
        'secret': '<insert enable secret here>'
 }
```

3. Next, we are going to use a Python context manager to open a connection, similar to opening a file. Inside of that context manager, we'll then use the `send_command()` method of the `ConnectHandler` to send our configuration commands as follows:

```
with ConnectHandler(**connection_info) as conn:
     conn.enable()
      out = conn.send_command("show running-config")
      print(out)
```

4. To run this script, you will have to go to your terminal and execute it. The output should be the same as if you had connected to the device by SSH directly and executed it with the following:

```
python3 enter_enable.py
```

How it works...

In this recipe, we are using netmiko's `ConnectHandler` to connect to our device using the connection information passed in from a dictionary that contains the `host`, `port`, `username`, `password`, and device type information required by netmiko.

Additionally, we have provided a secret. This is the secret used when trying to enter enable mode.

After establishing a connection, we prompt our netmiko connection to enter enable mode by using the `enable()` function on our `ConnectHandler`.

Authenticating using public-private keys with netmiko

In all recipes so far, we have used a combination of username and password to authenticate ourselves against the device. A different approach employed quite frequently, and sometimes even required by certain audit requirements, is that we use public-private keys to authenticate.

In *Chapter 2, Connecting to Network Devices via SSH Using Paramiko*, we saw that Paramiko supports the usage of keys instead of a username/password combination to authenticate. As netmiko is built on top of Paramiko, we can also do this conveniently with netmiko.

In this recipe, we'll see how to use a private key to authenticate against a device and issue a command.

Getting ready

Open your code editor and start by creating a file called `auth_key.py`. Next, navigate your terminal to the same directory that you just created the `auth_key.py` file in.

Make sure that the device you are using is configured to use public-private key authentication and that you have the correct private key.

How to do it...

We'll create a dictionary that contains our connection details. In this dictionary, contrary to what we have done so far, we won't specify a username/password configuration, but rather enable key authentication and then provide the path where our private key is stored for netmiko to use for authentication.

Follow these steps to authenticate using a specific key file instead of a username/password combination:

1. Import `ConnectHandler` from `netmiko`:

```
from netmiko import ConnectHandler
```

2. Create a dictionary that will contain our connection details. In this example, we are connecting to a Cisco IOS device. See the *How it works* section in the *Connect to a network device using netmiko* recipe for a list of device types from other vendors. Make sure to use the exact same dictionary keys (such as `device_type` or `host`) when specifying your connection information.

   Note how we are not providing a password:

```
connection_info = {
        'device_type': 'cisco_ios',
        'host': '<insert your host here>',
        'port': <insert your port number here>,
        'username': '<insert your username>',
        'use_keys': True,
        'key_file': '<insert path to private key here>'
}
```

3. Next, we are going to use a Python context manager for opening a connection, similar to how we open a file. Inside of that context manager, we'll then use the `send_command()` method of the `ConnectHandler` to send our `show interfaces` command to the device and print back the output:

```
with ConnectHandler(**connection_info) as conn:
        out = conn.send_command("show interfaces")
        print(out)
```

4. To run this script, go to your terminal and execute it with the following:

```
python3 auth_key.py
```

The output should be the same as if you had connected to the device by SSH directly and issued the command.

How it works...

In this recipe, we are providing a different set of inputs to the `ConnectHandler` for our connection. Instead of a password, we signal netmiko to use a private key for authentication by setting the `use_keys` flag to `True`. With key-based authentication enabled we then need to provide the path to our private key. This is done by specifying this path in the `key_file` argument.

With these changes to the connection details applied, we can proceed as seen before and create a `ConnectHandler` within a context manager that can send commands and retrieve the output.

Handling commands that prompt for information using netmiko

In the previous recipes in this chapter, we have always run commands that did not expect any user input. Once issued to the device, we could just wait for them to finish and see the output returned by netmiko. But what about commands that do prompt for confirmation? A typical example would be a `delete` command that can be used to delete a file from the device. This command will prompt for confirmation.

A naïve implementation could be to issue our `delete` command, wait a few moments to be sure that the device is showing the prompt, and then send the required confirmation, for example a line break or a `y` using another `send_command()` call. While this approach would work, it is clumsy and does not take into account changes to the required string. Since this is a common task, netmiko does provide the functionality to react to a command prompting for information.

In this recipe, we will issue a `delete` command and react to the prompt. While this example uses the `delete` command, any command that prompts for your input can be used.

Getting ready

Open your code editor and start by creating a file called `prompt.py`. Next, navigate your terminal to the same directory that you just created the `prompt.py` file in.

We will delete a file called `test_upload.txt` from the flash of our device in this example. You can upload such a test file programmatically by following the instructions in the *Copying files to a device using netmiko* recipe.

How to do it...

Let's start by first importing the required classes, specifying the connection details, and then establishing a connection to the device. With the connection to the device built, we can then issue our commands.

Follow these steps to answer the prompt of a command:

1. Import `ConnectHandler` from `netmiko`:

```
from netmiko import ConnectHandler
```

2. Create a dictionary that will contain our connection details. In this example, we are connecting to a Cisco IOS device. See the *How it works* section in the *Connect to a network device using netmiko* recipe for a list of device types from other vendors. Make sure to use the exact same dictionary keys (such as `device_type` or `host`) when specifying your connection information:

```
connection_info = {
        'device_type': 'cisco_ios',
        'host': '<insert your host here>',
        'port': <insert your port number here>,
        'username': '<insert your username>',
        'password': '<insert your password here>',
}
```

3. Next, we are going to use a Python context manager for opening a connection similar to opening a file. Inside of that context manager, we'll then use the `send_command()` method of the `ConnectHandler` to send our `delete` command to the device and expect and answer the prompts:

```
with ConnectHandler(**connection_info) as conn:
      conn.send_command("delete flash:/test_upload.txt",
                  expect_string=r"Delete filename",
                  strip_prompt=False,
                  strip_command=False)
      conn.send_command("\n",
                  expect_string=r"confirm",
                  strip_prompt=False,
                  strip_command=False)
     conn.send_command("y",
                  expect_string=r"#",
                  strip_prompt=False,
                  strip_command=False)
```

4. To run this script, go to your terminal and execute it with the following command:

```
python3 prompt.py
```

There should be no output but the file should be deleted on the target device.

How it works...

In this recipe, we are using netmiko's `ConnectHandler` to connect to our device using the connection information passed in from a dictionary that contains the host, port, username, password, and device type information required by netmiko.

With the connection established, we can use the `send_command` function to execute our command. We passed three additional keyword arguments (`expect_string`, `strip_prompt`, and `strip_command`) to our command. The `strip_prompt` and `strip_command` flags are turned to `false` to include the command issued, as well as the prompt (for example, #) in the output. The netmiko documentation, which you can find at `https://github.com/ktbyers/netmiko/blob/develop/EXAMPLES.md#handling-commands-that-prompt-timing`, specifies that this makes the output easier to read, and when sending our final `y` we do expect the prompt (in this example, #) to appear, indicating that the command has run successfully with all prompts accepted.

The interesting parameter is the `expect_string` parameter. Here we are passing a regular expression, marked by the `r` before the string, to our command. In the background, the session waits until a string matching the regular expression we passed is found in the output. Once a string matching the regular expression is found the command finishes executing. This allows us to wait until the device is actually ready and is prompting us for an input before sending the required two inputs, in this example an *Enter* keypress or linebreak and a `y` for confirmation. The exact sequence of commands, as well as the strings to expect, depend on the command you are issuing.

There's more...

In the preceding example, we used a new Python construct to define a regular expression called a **raw string**. In order to understand how raw strings differ from regular strings, we have to take a look at how Python normally handles the \ character. We have seen before that we can use `\n` to define a newline or use `\r` to define a carriage return. This means that Python, when using normal strings, treats the character following \ differently. Python tries to find a matching escape sequence, such as the newline for `\n` or `\t` for a *Tab* keypress. These sequences are also called escape sequences.

But what if instead of having \ signal the beginning of an escape sequence, we want the slash to just be treated like a slash? Especially when dealing with regular expressions, this can become a problem as we want the \ to be treated literally instead of as the start of an escape sequence, as in most regular expressions. We could escape the slash by using two slashes – that is, in order to have Python print a single slash, we have to write `print("\\")` instead of just `print("\")`. The former command will error out with a `SyntaxError: EOL while scanning string literal` error, while the latter will print out a single \. Always escaping the slashes can become a pain and thus Python offers us raw strings. These raw strings, signaled by the `r` in front of the quotation marks, tell Python to treat the following string without searching for and processing the string literals.

Have a look at the output of the following code sample:

```
print("\n")
print(r"\n")
```

While the first `print` just prints two new lines (one from the `\n` and one that is always printed with each `print` command), the second `print` command will print the `\n` as a literal string instead of treating it as a special sequence indicating a newline character.

5
Model-Driven Programmability with NETCONF and ncclient

In previous chapters, we learned how to connect to a network device and either retrieve or change the configuration of our device using the same CLI commands that we can also issue manually. Now, let's take a step back and examine what we are actually doing when we issue configuration commands. Fundamentally, a device has a certain state, which is the sum of all the configuration variables and their values. Every time we issue a command, we either retrieve this state or change the state by creating, modifying, or deleting variables within that state.

While CLI commands are a great way to modify states for humans, as they are easy to read and, therefore, easy to remember and type, they are not ideal for computers. In *Chapter 4, Configuring Network Devices Using Netmiko*, in the *Retrieving command outputs as structured Python data using netmiko and Genie* recipe, we learned that we need an entire library (for example, Genie) to take the human-readable CLI output and convert it into a machine-readable data format. This is a cumbersome task because we are also trying to teach the computer how to parse information that was primarily intended for consumption by humans. CLIs and their outputs were never *designed* to be consumed by computers; so, in this chapter, we will explore a format that was *made for* computers.

Of course, we are not the first ones to identify the shortcomings of previous methods. The gap in the standard way of retrieving and changing the configuration of network devices has also been identified by the **Internet Engineering Task Force** (**IETF**). In 2006, they published their approach for changing or retrieving configurations and, therefore, the state of a network device. They named their new protocol **Network Configuration Protocol** (**NETCONF**).

In this chapter, we will explore the different ways in which we can use the ncclient library to manipulate a device's state by using the NETCONF protocol. Specifically, we will be covering the following topics and recipes:

- Revisiting the NETCONF and YANG modules
- Connecting to a network device using ncclient
- Using NETCONF and ncclient to retrieve the running configuration
- Using NETCONF and ncclient to change the starting configuration
- Retrieving an interface configuration using NETCONF and ncclient
- Changing an interface configuration using NETCONF and ncclient
- Reacting to event notifications using NETCONF and ncclient

Technical requirements

For this chapter and the remainder of the book, you'll require an installation of Python. Specifically, you'll need a Python interpreter of version 3.6.1 or higher. This book makes use of the language constructs of Python 3 and, therefore, is incompatible with Python 2.x. Please also install the `ncclient` package (that is, `python3 -m pip install ncclient`). At the time of writing, we are using the latest version of `ncclient`, that is, version 0.6.10. You can install this exact version by issuing `python3 -m pip install ncclient==0.6.10`. All of the code examples have been developed and tested on a macOS X machine, running Mac OS X version 10.15.4.

You will also need a code editor. Popular choices include Microsoft Visual Studio Code or Notepad++. Additionally, you'll need a device (either virtual or physical) that you can use to log in via SSH.

You'll also need the login information of a networking device that supports the NETCONF protocol. By default, NETCONF uses TCP port `830` and SSH as its transport protocol. Please make sure that your firewall does not block any communications on this port. You will have to verify that your device supports NETCONF and that NETCONF is enabled. Please refer to the configuration guide of your device to verify that NETCONF is supported and to find out how to enable it.

You can view this chapter's code in action here: `https://bit.ly/2VPzqR1`

Revisiting the NETCONF and YANG modules

Before jumping straight into the Python code, let's revisit how NETCONF works and how we use YANG modules to describe the data model used to retrieve or modify our device configuration. At its core, NETCONF works by sending operations, which are described in **eXtended Markup Language** (**XML**), to a device. The device then carries out that operation and returns information.

The core operations defined by the NETCONF protocol are as follows:

- `<get>`: This is used to retrieve information about the state of a device.
- `<get-config>`: This is used to retrieve an entire configuration (for example, the entire running configuration of your device).
- `<edit-config>`: This is used to edit a configuration (for example, to change the entire running configuration on your device). Here, *edit* can refer to creating, deleting, or combining configuration variables.
- `<copy-config>`: This is used to copy from one configuration to another (for example, to copy your running configuration to your startup configuration).
- `<delete-config>`: This is used to delete an entire configuration.
- `<lock>`: This is used to lock a configuration (for example, your running configuration). This allows you to make sure that no other process is manipulating this specific configuration.
- `<unlock>`: This is used to release a lock that was previously obtained using the `<lock>` operation.
- `<close-session>`: This is used to terminate a NETCONF session.
- `<kill-session>`: This is used to force the termination of a NETCONF session.

So far, we have learned that the NETCONF module defines a set of operations that can be used to carry out all of the different operations required to programmatically retrieve or modify the configurational state of a network device. What NETCONF does not define (currently) is what this operation should look like. The content of a NETCONF request needs to be written in well-formed XML, but what should this XML look like?

This is where **Yet Another Next Generation** (**YANG**) comes into play. YANG is a modeling language that can be used to define what the data that is being sent over a network connection using NETCONF should look like. While YANG is independent of the protocol it is being used in, in this chapter, we will focus on NETCONF and, therefore, only look at XML-formatted messages. In *Chapter 8, Configuring Devices Using RESTCONF and requests*, we will examine how YANG modules can also be implemented using JSON-formatted messages.

Each YANG module is part of a namespace that defines the data types of the tree structure this YANG module is modeling for us. We do not only have individual items in our `leaf` instances but also lists of these items. Let's take a look at an example of a YANG module that describes a computer, namely, a computer's storage and compute resources:

```
module example-computer {
      namespace http://example.com/example-computer;
      prefix computer;

      container devices {
            config true;

            list storage {
                  key identifier;
                  leaf identifier {
                        type string; mandatory true;
                  }
                  leaf type {
                        type string; mandatory true;
                  }
            }
            typedef clock_rate {
                  type uint16;
                  description "The clock rate of a CPU in
hertz"
```

```
            }

            list compute {
                  key core;
                  leaf core { type uint16; mandatory true; }
                  leaf rate { type clock_rate; mandatory true;
}
            }
      }
}
```

First, we define a container called `devices`. This container will contain our lists of `leaf` instances. In our example, we are modeling a list of storage devices. Here, a storage device is defined by its identifier (`dev0`) and its type. For example, this could be information about whether this device is a hard drive or an SSD. Besides the default types (such as a string or `uint16`), YANG also allows us to define our own data types. In this example, we are defining a `clock_rate` data type that will model the clock rate of our next list and the `compute` resources of our device. Each of these resources is represented by a core number and a rate that has a custom data type, called `clock_rate`.

An example of an XML document that would be a valid representation of the model is shown in the following snippet:

```
<data xmlns="urn:ietf:params:xml:ns:netconf:base:1.0">
  <devices>
    <storage>
      <identifier>sd0</identifier>
      <type>SSD</type>
    </storage>
    <storage>
      <identifier>sd1</identifier>
      <type>HDD</type>
    </storage>
    <compute>
      <core>1</core>
      <rate>36000</rate>
    </compute>
  </devices>
</data>
```

Thus, YANG is a powerful tool that you can use to model how data should be structured. And while we could use YANG to model any kind of data, it is predominantly used to model the data of the configurational state of network devices.

So, to quickly summarize, the modeling language of YANG is used to describe how the content of a message that is being sent to a device capable of being configured via NETCONF needs to be formatted. This information includes what kind of fields are included, what fields can be changed by the user, and what fields are read-only.

You might be wondering who defines these modules and keeps them up to date. There are certain models provided by `OpenConfig`. `OpenConfig` takes common network structures, such as VLANs or the configuration of the **Open Shortest Path First** (**OSPF**) **protocol**, and implements them consistently across different vendors. The benefit of writing an automation workflow purely using the YANG modules of `OpenConfig` is that the same workflow can be applied to the devices of different vendors and will have the same configurational effect. This is extremely useful when you are dealing with multi-vendor networks.

For those features that are specific to a vendor platform, most vendors, including Huawei, Cisco, and Juniper, maintain platform- and vendor-specific YANG modules for their specific devices and features. `https://yangcatalog.org/` is a great resource for validating and exploring your YANG-based projects. Additionally, you can download all of the available YANG modules by cloning the GitHub repository at `https://github.com/YangModels/yang.git` for offline reference.

Connecting to a network device using ncclient

With the introduction to the NETCONF and YANG modules out of the way, we can now discuss how to connect to a network device using Python so that we can later issue a NETCONF command to it. NETCONF is built on top of the **Remote Procedure Call** (**RPC**) protocol, and while we could use a generic library that implements this protocol, the `ncclient` package has been purposely built to be used to connect to network devices using NETCONF and, therefore, issue RPC calls to the device.

In this example, we'll set up our connection details, open a connection, and then retrieve the list of all the YANG modules that are currently supported by the device we are connecting to. Then, we'll filter this list of available YANG modules to identify the `OpenConfig` modules and print the list of found modules back to the user.

Getting ready

Open your code editor and create a file called `connect.py`. Next, in your Terminal, navigate to the same directory that you just created the `connect.py` file in.

How to do it...

The following steps will allow us to connect to a network device and retrieve all of the available `OpenConfig` YANG modules:

1. Import the manager from `ncclient` using the following command:

```
from ncclient import manager
```

2. Create a dictionary with all of the connection details:

```
conn_info = {
        "host": "<insert your host here>",
    "port": <insert the port number here>,
    "username": "<insert the username here>",
    "password": "<insert the password here>",
    "hostkey_verify": False
}
```

3. With the connection details defined, we can create a connection to our network device using the `ncclient` manager, as follows:

```
device = manager.connect(**conn_info)
```

4. With our connection established, we can now retrieve the list of available modules. These are also known as the device's capabilities:

```
capabilities = device.server_capabilities
```

5. Finally, we can iterate over each of those capabilities and search for the YANG modules that are defined by `OpenConfig` as shown:

```
for cap in capabilities:
          if "http://openconfig.net/yang/" in cap:
         print(cap)
```

6. To run this script, go to your Terminal and execute it using the following command:

```
python3 connect.py
```

The output of this command should be a long list of `OpenConfig` YANG modules that all start with `http://openconfig.net/yang/`. The exact list of modules depends on the device you are connecting to:

```
→ ch05 git:(master) ✗ python3 connect.py
http://openconfig.net/yang/aaa?module=openconfig-aaa&revision=2017-09-18
http://openconfig.net/yang/aaa/types?module=openconfig-aaa-types&revision=2017-09-18
http://openconfig.net/yang/acl?module=openconfig-acl&revision=2017-05-26&deviations=cisco-xe-openconfig-acl-deviation
http://openconfig.net/yang/alarms?module=openconfig-alarms&revision=2017-08-24
http://openconfig.net/yang/bgp?module=openconfig-bgp&revision=2016-06-21
http://openconfig.net/yang/bgp-policy?module=openconfig-bgp-policy&revision=2016-06-21&deviations=cisco-xe-openconfig-bgp-policy-deviation
http://openconfig.net/yang/bgp-types?module=openconfig-bgp-types&revision=2016-06-21
http://openconfig.net/yang/cisco-xe-openconfig-if-ethernet-deviation?module=cisco-xe-openconfig-if-ethernet-deviation&revision=2017-11-01
http://openconfig.net/yang/cisco-xe-openconfig-if-ip-deviation?module=cisco-xe-openconfig-if-ip-deviation&revision=2017-03-04
http://openconfig.net/yang/cisco-xe-openconfig-interfaces-deviation?module=cisco-xe-openconfig-interfaces-deviation&revision=2018-03-27
http://openconfig.net/yang/cisco-xe-openconfig-system-deviation?module=cisco-xe-openconfig-system-deviation&revision=2017-11-27
http://openconfig.net/yang/header-fields?module=openconfig-packet-match&revision=2017-05-26
http://openconfig.net/yang/interfaces?module=openconfig-interfaces&revision=2018-01-05&deviations=cisco-xe-openconfig-interfaces-deviation
http://openconfig.net/yang/interfaces/aggregate?module=openconfig-if-aggregate&revision=2018-01-05
http://openconfig.net/yang/interfaces/ethernet?module=openconfig-if-ethernet&revision=2018-01-05&deviations=cisco-xe-openconfig-if-ethernet-dev
iation
http://openconfig.net/yang/interfaces/ip?module=openconfig-if-ip&revision=2018-01-05&deviations=cisco-xe-openconfig-if-ip-deviation,cisco-xe-op
enconfig-interfaces-deviation
http://openconfig.net/yang/interfaces/ip-ext?module=openconfig-if-ip-ext&revision=2018-01-05
http://openconfig.net/yang/local-routing?module=openconfig-local-routing&revision=2016-05-11
http://openconfig.net/yang/network-instance?module=openconfig-network-instance&revision=2017-01-13&deviations=cisco-xe-openconfig-bgp-deviation
,cisco-xe-openconfig-network-instance-deviation
http://openconfig.net/yang/network-instance-l3?module=openconfig-network-instance-l3&revision=2017-01-13
http://openconfig.net/yang/network-instance-types?module=openconfig-network-instance-types&revision=2016-12-15
http://openconfig.net/yang/openconfig-ext?module=openconfig-extensions&revision=2017-04-11
http://openconfig.net/yang/openconfig-types?module=openconfig-types&revision=2018-01-16
http://openconfig.net/yang/packet-match-types?module=openconfig-packet-match-types&revision=2017-05-26
http://openconfig.net/yang/platform?module=openconfig-platform&revision=2016-12-22
http://openconfig.net/yang/platform-types?module=openconfig-platform-types&revision=2017-08-16
http://openconfig.net/yang/platform/port?module=openconfig-platform-port&revision=2016-10-24
http://openconfig.net/yang/platform/transceiver?module=openconfig-platform-transceiver&revision=2017-09-18
http://openconfig.net/yang/policy-types?module=openconfig-policy-types&revision=2016-05-12
http://openconfig.net/yang/rib/bgp?module=openconfig-rib-bgp&revision=2017-03-07
http://openconfig.net/yang/rib/bgp-ext?module=openconfig-rib-bgp-ext&revision=2016-04-11
http://openconfig.net/yang/rib/bgp-types?module=openconfig-rib-bgp-types&revision=2016-04-11
http://openconfig.net/yang/routing-policy?module=openconfig-routing-policy&revision=2016-05-12&deviations=cisco-xe-openconfig-routing-policy-de
viation
http://openconfig.net/yang/system?module=openconfig-system&revision=2018-01-21&deviations=cisco-xe-openconfig-system-deviation
http://openconfig.net/yang/system/logging?module=openconfig-system-logging&revision=2017-09-18
```

Figure 5.1 – The output of the available modules as they are retrieved by our script

The preceding screenshot shows an excerpt of the scripts output when connecting to a Cisco IOS XE device, which is running IOS version 16.9.3.

How it works...

First, we import the manager from `ncclient`. This manager serves as the central point for all our interactions with the network device. Next, we create a dictionary that contains the connection details for our device, including the following:

- The host on which to connect
- The username we will use to connect
- The password to use when connecting
- The disabling of the host key verification

Then, we use variadic arguments to pass all of the connection variables that were defined in the dictionary to our manager. Please refer to the *There's more...* section of the *Connecting to a network device using netmiko* recipe, in *Chapter 4*, *Configuring Network Devices Using Netmiko*, for more information regarding variadic arguments and how they get passed to a function in Python.

Once our connection has been established, we can use the manager's `capabilities` variable to retrieve a list of all the different YANG modules that are available on this specific device. With our available YANG modules or capabilities saved to a list, we can then loop over each of them. The capabilities list is a list of strings that represent the names of the YANG modules. As we learned in the introduction to YANG modules (please refer to the *Revisiting the NETCONF and YANG modules* section), each of these modules has a namespace. Since we are interested in those that are part of the `OpenConfig` module, we check whether that particular namespace (`http://openconfig.net/yang`) is part of our YANG module's name. Then, we use an if-clause together with the `in` statement to check whether our prefix is part of the capability.

Using NETCONF and ncclient to retrieve the running configuration

So far, we have only established a connection to our device, but we have not actually retrieved any configuration data. This is about to change in this recipe. Using NETCONF and `ncclient`, we are now going to establish a connection to our network device and retrieve the currently running configuration. This configuration will be in the form of an XML file that represents the running configuration of our device according to the format defined in the YANG modules for our device.

Getting ready

Open your code editor and create a file called `get_running.py`. Next, in your Terminal, navigate to the same directory that you just created the `get_running.py` file in.

You'll also need the login information of a networking device that supports the NETCONF protocol.

How to do it...

To establish a connection to your device and retrieve the currently running configuration in XML format, follow these steps:

1. Import the manager from `ncclient` along with the `prettyprint` module from the standard library:

```
from ncclient import manager
import pprint
```

2. Create a dictionary with all of the connection details:

```
conn_info = {
        "host": "<insert your host here>",
  "port": <insert the port number here>,
  "username": "<insert the username here>",
  "password": "<insert the password here>",
  "hostkey_verify": False
}
```

3. With the connection details defined, we can create a connection to our network device using the `ncclient` manager, as follows:

```
device = manager.connect(**conn_info)
```

4. With our connection established, we can retrieve the running configuration using the following:

```
conf = device.get_config(source='running').data_xml
```

5. Finally, print out the running configuration's XML:

```
pprint.pprint(conf)
```

6. To run this script, go to your Terminal and execute it using the following command:

```
python3 get_running.py
```

The output should be a long XML document representing your running configuration:

Figure 5.2 – Our running configuration in XML format

The preceding screenshot shows an excerpt of running our configuration retrieval script against a Cisco IOS XE device, which is running IOS version 16.9.3.

How it works...

First, we import the manager from `ncclient`. This manager serves as the central point for all our interactions with the network device. Next, we create a dictionary that contains the connection details for our device. Please refer to the *Connecting to a network device using ncclient* recipe for a more detailed explanation of the connection process.

With our connection established, we can use the manager's built-in `get_config()` method to signal that we want to invoke a NETCONF `<get-config>` operation. We need to specify the source, that is, which configuration store on the device to draw from, as the only argument to `get_config()`. In our case, we want to retrieve the running configuration and, therefore, use `running`. `get_config()` returns an object that has the `data_xml` property, which is the XML representation of the running configuration that we are after. We can then print out the configuration we just retrieved using `prettyprint` to get an output that is formatted so that it is easier to read.

There's more...

While the core concept of ncclient is to be vendor agnostic, version 0.4.1 introduced a number of changes that allowed for more vendor-specific handling. This means that if you know the device you are running against, you can give this information to ncclient and the library will adapt to the specifics of the devices operating system/platform.

To do so, we need another connection detail, called device_params, which is a dictionary that contains device-specific information such as the name of this device/vendor. We can modify our conn_info dictionary to look like this when we are dealing with a Juniper device:

```
conn_info = {
        "host": "<insert your host here>",
    "port": <insert the port number here>,
    "username": "<insert the username here>",
    "password": "<insert the password here>",
    "hostkey_verify": False,
    "device_params": {
         "name": "junos"
     }
}
```

Notice that the additional device_params parameter is a dictionary itself. At the time of writing, the following device handlers and their associated vendors are supported:

- junos for Juniper
- csr for Cisco CSR devices
- nexus for Cisco Nexus devices
- iosxr for Cisco IOS XR devices
- iosxe for Cisco IOS XE devices
- huawei or huaweiyang for Huawei devices
- alu for Alcatel Lucent devices
- h3c for H3C devices
- hpcomware for HP Comware devices
- default for anything that you either don't know the vendor of or is not part of the preceding list

Using NETCONF and ncclient to change the starting configuration

Retrieving the running configuration, as we will learn in the following recipes (please refer to the *Changing an interface configuration using NETCONF and ncclient* recipe), is already a powerful thing to do. But what if we want to overwrite a configuration? Perhaps we want to provide a completely new configuration from a local file or overwrite the starting configuration with our currently running configuration. This is what we are going to explore in this recipe. We'll learn how to use the `copy_config()` function from `ncclient` to copy the currently running configuration to a file and then copy it over to the starting configuration.

Getting ready

Open your code editor and create a file called `change_starting.py`. Next, in your Terminal, navigate to the same directory that you just created the `change_starting.py` file in.

You'll also need the login information of a networking device that supports the NETCONF protocol.

How to do it...

To, first, copy a running configuration to a file and then copy it to the starting configuration, follow these steps:

1. Import the manager from `ncclient` using the following command:

```
from ncclient import manager
```

2. Create a dictionary with all of the connection details:

```
conn_info = {
        "host": "<insert your host here>",
    "port": <insert the port number here>,
    "username": "<insert the username here>",
    "password": "<insert the password here>",
    "hostkey_verify": False
}
```

3. With the connection details defined, we can create a connection to our network device using the `ncclient` manager, as follows:

```
device = manager.connect(**conn_info)
```

4. With our connection established, we can first copy the running configuration to a new file, as shown here:

```
device.copy_config(source='running', target='file:///
current_running.conf')
```

5. Next, we can copy this newly created `current_running.conf` file over to our starting configuration using the following:

```
device.copy_config(source='file:///current_running.conf',
target='starting')
```

6. To run this script, go to your Terminal and execute it using the following command:

```
python3 change_starting.py
```

After running this script, you should see that your starting configuration has been replaced with your running configuration. You can also use this workflow to create configuration backups and allow for automated rollbacks if one of your changes goes wrong.

How it works...

First, we import the manager from `ncclient`. This manager serves as the central point for all our interactions with the network device. Next, we create a dictionary that contains the connection details for our device. Please refer to the *Connecting to a network device using ncclient* recipe for a more detailed explanation of the connection process.

After we have established our device connection, we copy our currently running configuration over to a file called `current_running.conf`. With this operation complete, we then use the same `copy_config()` function of the `ncclient` manager to copy the configuration from our newly created file to our starting configuration.

Retrieving an interface configuration using NETCONF and ncclient

In the previous recipes (for example, the *Using NETCONF and ncclient to retrieve the running configuration* recipe), we have always dealt with the entire running configuration of our device. While this can be useful in some cases, often, we want to focus on one aspect of our network device such as the list of interfaces. NETCONF has support for this through a concept called filters. Filters allow us to specify criteria on which to filter. Following this, NETCONF will only return entries that match the filter criterion. Filters in NETCONF are, as with everything, defined by an XML document and can be passed on to `ncclient` to be sent to the device.

In this recipe, we will examine how to filter our running configuration to only return the interface configuration.

Getting ready

Open your code editor and create a file called `get_interface.py`. Next, in your Terminal, navigate to the same directory that you just created the `get_interface.py` file in.

You'll also need the login information of a networking device that supports the NETCONF protocol.

How to do it...

To retrieve the configured interfaces of your device, follow these steps:

1. Import the manager from `ncclient` using the following command:

```
from ncclient import manager
import pprint
```

2. Create a dictionary with all of the connection details, as follows:

```
conn_info = {
    "host": "<insert your host here>",
    "port": <insert the port number here>,
    "username": "<insert the username here>",
    "password": "<insert the password here>",
    "hostkey_verify": False
}
```

3. With the connection details defined, we can create a connection to our network device with the `ncclient` manager using the following commands:

```
device = manager.connect(**conn_info)
```

4. With our connection established, we can now define our filter:

```
filter = """
<filter>
  <interfaces xmlns="urn:ietf:params:xml:ns:yang:ietf-
interfaces">
    <interface>
    </interface>
  </interfaces>
</filter>
"""
```

5. With our filter defined, we can apply it to our `get_config()` command, which is targeted at the running configuration:

```
conf = device.get_config(source='running',
filter=filter).data_xml
```

6. Finally, print out the XML of the interfaces that was defined in our running configuration:

```
pprint.pprint(conf)
```

7. To run this script, go to your Terminal and execute it using the following command:

```
python3 get_interface.py
```

The output should be a long XML document that represents the interface data of your running configuration:

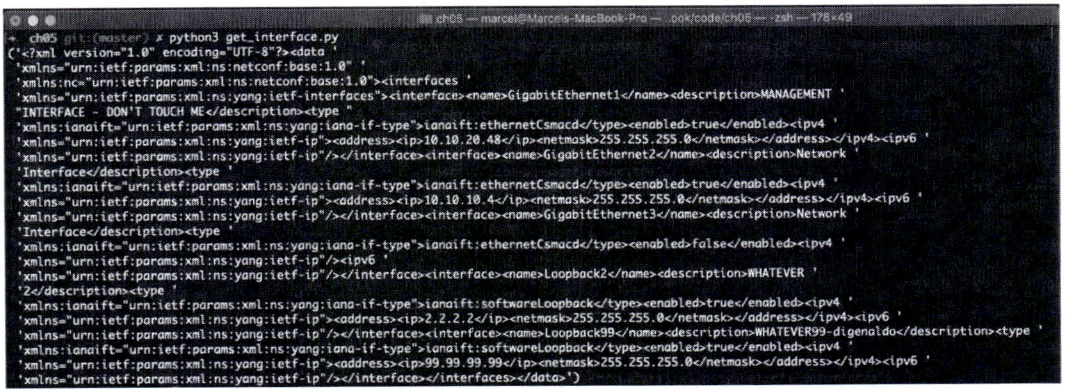

```
ch05 — marcel@Marcels-MacBook-Pro — ..ook/code/ch05 — -zsh — 178×49
→ ch05 git:(master) ✗ python3 get_interface.py
('<?xml version="1.0" encoding="UTF-8"?><data '
 'xmlns="urn:ietf:params:xml:ns:netconf:base:1.0" '
 'xmlns:nc="urn:ietf:params:xml:ns:netconf:base:1.0"><interfaces '
 'xmlns="urn:ietf:params:xml:ns:yang:ietf-interfaces"><interface><name>GigabitEthernet1</name><description>MANAGEMENT '
 "INTERFACE - DON'T TOUCH ME</description><type "
 'xmlns:ianaift="urn:ietf:params:xml:ns:yang:iana-if-type">ianaift:ethernetCsmacd</type><enabled>true</enabled><ipv4 '
 'xmlns="urn:ietf:params:xml:ns:yang:ietf-ip"><address><ip>10.10.20.48</ip><netmask>255.255.255.0</netmask></address></ipv4><ipv6 '
 'xmlns="urn:ietf:params:xml:ns:yang:ietf-ip"/></interface><interface><name>GigabitEthernet2</name><description>Network '
 'Interface</description><type '
 'xmlns:ianaift="urn:ietf:params:xml:ns:yang:iana-if-type">ianaift:ethernetCsmacd</type><enabled>true</enabled><ipv4 '
 'xmlns="urn:ietf:params:xml:ns:yang:ietf-ip"><address><ip>10.10.10.4</ip><netmask>255.255.255.0</netmask></address></ipv4><ipv6 '
 'xmlns="urn:ietf:params:xml:ns:yang:ietf-ip"/></interface><interface><name>GigabitEthernet3</name><description>Network '
 'Interface</description><type '
 'xmlns:ianaift="urn:ietf:params:xml:ns:yang:iana-if-type">ianaift:ethernetCsmacd</type><enabled>false</enabled><ipv4 '
 'xmlns="urn:ietf:params:xml:ns:yang:ietf-ip"/><ipv6 '
 'xmlns="urn:ietf:params:xml:ns:yang:ietf-ip"/></interface><interface><name>Loopback2</name><description>WHATEVER '
 '2</description><type '
 'xmlns:ianaift="urn:ietf:params:xml:ns:yang:iana-if-type">ianaift:softwareLoopback</type><enabled>true</enabled><ipv4 '
 'xmlns="urn:ietf:params:xml:ns:yang:ietf-ip"><address><ip>2.2.2.2</ip><netmask>255.255.255.0</netmask></address></ipv4><ipv6 '
 'xmlns="urn:ietf:params:xml:ns:yang:ietf-ip"/></interface><interface><name>Loopback99</name><description>WHATEVER99-digenaldo</description><type '
 'xmlns:ianaift="urn:ietf:params:xml:ns:yang:iana-if-type">ianaift:softwareLoopback</type><enabled>true</enabled><ipv4 '
 'xmlns="urn:ietf:params:xml:ns:yang:ietf-ip"><address><ip>99.99.99.99</ip><netmask>255.255.255.0</netmask></address></ipv4><ipv6 '
 'xmlns="urn:ietf:params:xml:ns:yang:ietf-ip"/></interface></interfaces></data>')
```

Figure 5.3 – Our interface list in XML form

The preceding screenshot shows our interface retrieval script running against a Cisco IOS XE device, which is running IOS version 16.9.3.

How it works...

First, we import the manager from `ncclient`. This manager serves as the central point for all our interactions with the network device. Next, we create a dictionary that contains the connection details for our device. Please refer to the *Connecting to a network device using ncclient* recipe for a more detailed explanation of the connection process.

With our connection established, we define our filter. Filters are within the `<filter>` parent element within which we need to define what we want to retrieve. In our case, we are interested in all of the elements matching the IETF YANG module, `ietf-interfaces`. We can then pass this filter along to the `get_config()` function as the filter parameter. Finally, we can print out the resulting information about our interfaces back to us using the `prettyprint` library to get a better-formatted output.

Changing an interface configuration using NETCONF and ncclient

So far, we have learned how to retrieve or change our entire configuration. But what if we only want to modify certain aspects of our device's state? That is what we are going to cover in this recipe. As an example, we will use an `OpenConfig` YANG module to create a VLAN interface. While the code is specific to this example, in the sense that we are generating XML that can be used to create a VLAN, the workflow itself is always the same:

1. Generate an XML message that is formatted according to the specifications of the YANG module.
2. Send the generated XML to your network device using `ncclient` and the `edit_config()` method by choosing the running configuration as your target. This is similar to how we used the running configuration as our source in the `get_config()` call.

Getting ready

Open your code editor and create a file called `create_vlan.py`. Next, in your Terminal, navigate to the same directory that you just created the `create_vlan.py` file in.

You'll also need the login information of a networking device that supports the NETCONF protocol.

How to do it...

To create the necessary XML for a new VLAN and to send the newly created VLAN configuration to the networking device, follow these steps:

1. Import the manager from `ncclient` using the following command:

```
from ncclient import manager
```

2. Create a dictionary with all of the connection details using the following commands:

```
conn_info = {
        "host": "<insert your host here>",
    "port": <insert the port number here>,
    "username": "<insert the username here>",
    "password": "<insert the password here>",
    "hostkey_verify": False
}
```

3. With the connection details defined, we can create a connection to our network device using the `ncclient` manager:

```
device = manager.connect(**conn_info)
```

4. With our connection established, we can now create the XML that we want to send to our device:

```
vlan_conf = """
<vlan xmlns=http://openconfig.net/yang/vlan>
            <vlan-id>10</vlan-id>
            <config>
                  <name>New VLAN</name>
                  <status>ACTIVE</status>
                  <vlan-id>10</vlan-id>
            </config>
</vlan>
"""
```

5. With our XML ready, we can now send it to the device by using the `edit_config()` method of the manager:

```
device.edit_config(target="running", config=vlan_conf)
```

6. To run this script, go to your Terminal and execute it using the following command:

```
python3 create_vlan.py
```

Once this script has run, your device should have a new VLAN called `New VLAN` with an ID of `10`.

How it works...

First, we import the manager from `ncclient`. This manager serves as the central point for all our interactions with the network device. Next, we create a dictionary that contains all the connection details for our device. Please refer to the *Connecting to a network device using ncclient* recipe for a more detailed explanation of the connection process.

With our connection established, we can then create the necessary XML for a new VLAN. In this case, we are not using a vendor-specific YANG module, but rather, we are using the `OpenConfig` YANG module. You can check whether this specific module is supported on your device by searching the capabilities list of your device. Please refer to the *Connecting to a network device using ncclient* recipe to learn how to retrieve the list of `OpenConfig` YAML modules that are supported on the device you are using. In the XML, we specify all the information we want to configure on our VLAN, such as the name and the ID.

With the XML ready to be sent, we can use the manager's `edit_config()` method to actually send the changes we want to apply and that we have defined in our XML to the device. Note that `ncclient` has modeled its function names after those of the NETCONF protocol. We also need to specify a target. In our case, we want to overwrite the running configuration; therefore, we pass `running` as the argument to the target parameter.

Reacting to event notifications using NETCONF and ncclient

A common workflow task when dealing with network devices is the need to carry out a certain workflow whenever something of interest happens to our device. This "something of interest" could be a configuration change or a fault. NETCONF specifies a way for these events to be captured. Every time one of these events occurs, we can get a notification and act accordingly. One example could be to listen for events describing any sort of configuration change and then sending a message to a central chat room. In this recipe, you'll learn how to use `ncclient` to connect to a network device, subscribe to all notifications, and print out the contents of one notification if something happens.

Getting ready

Open your code editor and create a file called `get_notifications.py`. Next, in your Terminal, navigate to the same directory that you just created the `get_notifications.py` file in.

You'll also need the login information of a networking device that supports the NETCONF protocol.

How to do it...

To subscribe to all notifications from your device, follow these steps:

1. Import the manager from `ncclient` using the following command:

```
from ncclient import manager
```

2. Create a dictionary with all of the connection details using the following commands:

```
conn_info = {
        "host": "<insert your host here>",
    "port": <insert the port number here>,
    "username": "<insert the username here>",
    "password": "<insert the password here>",
    "hostkey_verify": False
}
```

3. With the connection details defined, we can create a connection to our network device using the `ncclient` manager:

```
device = manager.connect(**conn_info)
```

4. With our connection established, first, we need to tell `ncclient` to subscribe to the events, as follows:

```
device.create_subscription()
```

5. With our events subscribed to, we now have to wait until an event occurs. To do so, we have to go into an infinite loop to keep our connection up:

```
while True:
    notification = device.take_notification()
    print(notification.notification_xml)
```

6. To run this script, go to your Terminal and execute it using the following command:

```
python3 get_notifications.py
```

Your script should listen for notifications indefinitely, and you can trigger a notification by either running one of the scripts that will change the configuration or by manually changing the configuration from the CLI.

How it works...

First, we import the manager from `ncclient`. This manager serves as the central point for all of our interactions with the network device. Next, we create a dictionary that contains the connection details for our device. Please refer to the *Connecting to a network device using ncclient* recipe for a more detailed explanation of the connection process.

With our connection established, we can go ahead and create the subscriptions. The `create_subscrptions()` method, without any arguments, tells the manager to listen for notifications on all events. Once an event has been received, it is put in an internal queue. To retrieve messages from this queue, we use the `take_notification()` method, which listens on the internal notification queue and, as soon as the manager has put a new notification on it, will execute. Then, we print out the contents of our notification. Again, the notification is an XML-formatted message. With the infinite loop, we can go back to listen for a new notification once the previous notification has been processed/printed out.

6
Automating Complex Multi-Vendor Networks with NAPALM

When we consider the landscape of a modern networking infrastructure, one thing quickly becomes apparent: we are almost never dealing with a greenfield deployment. Equipment from different vendors gets upgraded and slotted into continuous cycles. Almost no network is powered purely by the solutions of one vendor, and even if you are doing a completely new greenfield deployment, most companies opt for multi-vendor solutions. However, this poses a problem for our network automation efforts. Without one single vendor to provide the equipment for our infrastructure, we don't have one common operating system running on the devices themselves. Therefore, due to the different operating systems of our devices, we have to write different scripts that interact with our devices differently.

In *Chapter 5, Model-Driven Programmability with NETCONF and ncclient*, we discussed one approach to solve this problem. Here, the devices were all supporting a common protocol (NETCONF) and type of data model to allow for vendor-agnostic automation. The topic of this chapter, NAPALM, takes a different approach to solve this problem. Instead of using a vendor-agnostic protocol on the device, NAPALM implements common functions within its library. In the background, these functions then execute differently depending on what type of device or vendor they are connected to. Therefore, we have a common Python interface that can deal with devices from different vendors. However, NAPALM does not stop there. Powered by this unified interface, NAPALM allows us to easily change, differentiate (or "diff"), and then apply or revert configuration changes across devices as well as verify the status of devices.

In this chapter, we will examine how to use NAPALM. In particular, we are going to use NAPALM for the following recipes:

- Connecting to devices from different vendors using NAPALM
- Issuing commands to a device using NAPALM
- Testing network reachability using ping and NAPALM
- Backing up your device configuration using NAPALM
- Gathering facts about your device using NAPALM
- Creating and applying a configuration template with jinja2 and NAPALM
- Rolling back configuration changes using NAPALM
- Validating your deployment using NAPALM

Technical requirements

For this chapter and the remainder of the book, you'll require an installation of Python. Specifically, you'll need a Python interpreter of version 3.6.1 or higher. This book makes use of the language constructs of Python 3 and, therefore, is incompatible with Python 2.x. Please also install the `napalm` package (that is, `python3 -m pip install napalm`). At the time of writing, we are using the latest version of `napalm`, which is version 3.2.0. You can install this exact version by issuing `python3 -m pip install ncclient==3.2.0`. All of the code examples have been developed and tested on a Mac OS X machine that is running Mac OS X version 10.15.4.

You will also need a code editor. Some popular choices include Microsoft Visual Studio Code or Notepad++. Additionally, you'll need a device (either virtual or physical) that you can log in to via SSH.

You can view this chapter's code in action here: `https://bit.ly/3iLeyU1`

Connecting to devices from different vendors using NAPALM

Let's start by learning how to connect to devices from different vendors using NAPALM. In *Chapter 4, Configuring Network Devices Using Netmiko*, we saw the idea of passing the type of device to the library so that the background code of our library can behave differently. NAPALM takes this concept to the next level and uses drivers to abstract the different vendors and provide a common interface.

Getting ready

Open your code editor and create a file called `connect_device.py`. Next, in your Terminal, navigate to the same directory that you just created the `connect_device.py` file in.

How to do it...

We will use the following steps to establish a connection to our network device using NAPALM:

1. Import the `napalm` module:

```
import napalm
```

2. From the `napalm` module, retrieve the appropriate driver for your device's vendor. Please refer to the *How it works...* section for a list of possible drivers:

```
driver = napalm.get_network_driver("<insert driver
name>")
```

3. Specify the connection details of your device. Note that you can pass a non-standard port in the optional arguments. If your device connects on the standard port, you can omit this:

```
conn_details = {
            "hostname": "<insert hostname>",
            "username": "<insert username>",
            "password": "<insert password>",
            "optional_args": {
                  "port": <insert port as integer>
            }
}
```

4. With the driver and connection details specified, we can create a new device:

```
device = driver(**conn_details)
```

5. Now we can establish a connection to the device, print out a success message, and then close the connection again:

```
device.open()
print("Succesfully connected to the device")
device.close()
```

6. To run this script, go to your Terminal and execute it using the following command:

```
python3 connect_device.py
```

How it works...

As mentioned in the introduction to this recipe, NAPALM uses the concept of drivers. These drivers abstract away the vendor-specific behavior of the device and, therefore, allow for a unified interface to be presented to the user.

The drivers for the different vendors are as follows:

- `eos` for EOS
- `junos` for Juniper
- `iosxr` and `ios` for the two different iOS variants from Cisco
- `nxos` and `nxos_ssh` for the Nexus switches from Cisco

While NAPALM *tries* to support all the functionality on all the different drivers, and therefore different vendors, some features are only available with certain drivers/devices. When designing your automation script, you can consult the support matrix provided by NAPALM. This matrix is available in the official documentation. Currently, it can be found at `https://napalm.readthedocs.io/en/latest/support/index.html`.

With the correct driver selected, we can then specify our device's connection details and then, using the driver, create a new device object. With this device object, we can then attempt to connect to our device using the `open()` method, print out a message that we have connected successfully, and then close the device connection again.

In this example, we have passed the username and password as variables in the script. While this is great for testing purposes, you can find a more secure way of passing on these variables in the *There's more...* section of this recipe.

There's more...

In this example, we passed a number of connection details such as the username, hostname, and password directly as variables in the script. While this is great for testing purposes, you might want to have your script prompt you for a variable upon execution. For non-secret variables, such as the username, host, and port, we can use the built-in `input()` function. However, for passwords, it's better to use a dedicated password prompt that hides what you have typed so that someone looking over your console history can't retrieve your password. For this purpose, Python has the built-in `getpass` module.

To retrieve the necessary configuration variables, not as static information in the script but rather, interactively from the user using a combination of input and the `getpass` module, follow these steps:

```
import getpass
SSH_PASSWORD = getpass.getpass(prompt='Password: ',
stream=None)
SSH_USER = input("Username: ")
SSH_HOST = input("Host: ")
SSH_PORT = int(input("Port: "))
```

Issuing commands to a device using NAPALM

With our connection to the device established, we can go ahead and issue a command and retrieve the output of the said command as text. This is similar to what we have previously done, first with paramiko in *Chapter 2, Connecting to Network Devices via SSH Using Paramiko*, and second with netmiko in *Chapter 4, Configuring Network Devices Using Netmiko*. NAPALM offers a very convenient method to specify multiple commands at once and retrieve all the output in one dictionary in a nicely organized way. If your device does have a driver for NAPALM available, then this is, by far, the most convenient method to retrieve the output out of paramiko, netmiko, and NAPALM.

Getting ready

Open your code editor and create a file called `send_command.py`. Next, in your Terminal, navigate to the same directory that you just created the `send_command.py` file in.

How to do it...

We will use the following steps to establish a connection to our network device and issue a list of commands to our device using NAPALM:

1. Import the `napalm` module:

```
import napalm
```

2. From the `napalm` module, retrieve the appropriate driver for your device's vendor. Please refer to the *How it works...* section of the *Connecting to devices from different vendors using NAPALM* recipe for a list of possible drivers:

```
driver = napalm.get_network_driver("<insert driver
name>")
```

3. Specify the connection details of your device. Note that you can pass a non-standard port in the optional arguments. If your device connects on the standard port, you can omit this:

```
conn_details = {
        "hostname": "<insert hostname>",
        "username": "<insert username>",
        "password": "<insert password>",
        "optional_args": {
                "port": <insert port as integer>
        }
}
```

4. With the driver and connection details specified, we can create a new device:

```
device = driver(**conn_details)
```

5. Specify the commands that we want to run on our device. In this example, we are going to retrieve the running configuration and the version:

```
commands = [
        "show running-configuration",
        "show version"
]
```

6. Open a connection to the device:

```
device.open()
```

7. With our device connected, we will issue the commands and store the results in a variable called `out`:

```
out = device.cli(commands)
```

8. Finally, print out the command's output and close the connection to the device:

```
print(out)
device.close()
```

9. To run this script, go to your Terminal and execute it using the following command:

```
python3 send_commands.py
```

The output should be a dictionary that contains, as the key, the commands (in our example, `show running-configuration` and `show version`) and, as the value for each command, the output from the device.

How it works...

First, we establish a connection to our device by specifying the connection details and retrieving the correct driver for our vendor. Please refer to the *Connecting to devices from different vendors using NAPALM* recipe for a more detailed explanation of how to establish a connection to a device using NAPALM.

With our connection established, we can specify the list of commands that we want to run. This list is passed on to the `cli()` method that has been provided by our device. This function takes the list of commands that we want to issue, executes them on the device, retrieves the output for each command, and then returns a dictionary where the key is the command we issued, and the value is the output of the command. This is a very convenient way of executing commands and retrieving their output.

In this example, we have passed the username and password as variables in the script. While this is great for testing purposes, you can find a more secure way of passing on these variables in the *There's more...* section of the *Connecting to devices from different vendors using NAPALM* recipe.

Testing network reachability using ping and NAPALM

So far, in this chapter, we have only discussed how to connect to a device (please refer to the *Connecting to devices from different vendors using NAPALM* recipe) and then how to issue a command (please refer to the *Issuing commands to a device using NAPALM* recipe). What we have not looked at, so far, are the vendor-agnostic methods that we promised.

The device we obtain from the vendor-specific driver offers a variety of functions; so, in this example, we are going to use the `ping()` function to test that our network device can reach a list of hosts.

Getting ready

Open your code editor and create a file called `ping.py`. Next, in your Terminal, navigate to the same directory that you just created the `ping.py` file in.

How to do it...

We will use the following steps to establish a connection with our network device. Then, we will use the `ping()` method to test connectivity:

1. Import the `napalm` module:

```
import napalm
```

2. From the `napalm` module, retrieve the appropriate driver for your device's vendor. Please refer to the *How it works...* section of the *Connecting to devices from different vendors using NAPALM* recipe for a list of possible drivers:

```
driver = napalm.get_network_driver("<insert driver
name>")
```

3. Specify the connection details of your device. Note that you can pass a non-standard port in the optional arguments. If your device connects on the standard port, you can omit this:

```
conn_details = {
      "hostname": "<insert hostname>",
      "username": "<insert username>",
      "password": "<insert password>",
      "optional_args": {
```

```
                "port": <insert port as integer>
        }
}
```

4. With the driver and connection details specified, we can create a new device:

```
device = driver(**conn_details)
```

5. Open a connection to the device:

```
device.open()
```

6. With our device connected, we can specify the hosts that we want to ping. In this example, I am specifying one host that will be pingable and one that won't be pingable in order to view the different outputs:

```
to_ping = [
        "packtpub.com",
        "10.0.0.1"
]
```

7. Next, we loop over each of the hosts and ping them using the device's `ping()` method. Then, we close the device connection once all hosts have been pinged:

```
for host in to_ping:
        out = device.ping(host)
        print(f"Results for {host}")
        print(out)
device.close()
```

8. To run this script, go to your Terminal and execute it using the following command:

```
python3 ping.py
```

The output should be two dictionaries. Please refer to the *How it works...* section for more details on what kind of information should be present.

How it works...

First, we establish a connection to our device by specifying the connection details and retrieving the correct driver for our vendor. Please refer to the *Connecting to devices from different vendors using NAPALM* recipe for a more detailed explanation of how to establish a connection to a device using NAPALM.

With our connection established and the hosts we want to ping defined, we can go ahead and use our first vendor-agnostic function, `ping()`, to ping the specified hosts. As you can see, we are not using the `cli()` method to issue a ping command to our device, and the output we retrieve is not plain text, which we'd have to parse. The output we get is a dictionary with two different possible keys. If the ping fails, the dictionary will contain a key, called `error`, and the value associated with that key will be the error message returned by ping. In the case of a successful ping, the dictionary contains a key called `success`, which itself maps to another dictionary. Contained in this dictionary are, among other things, the number of probes sent (`probes_sent`), the packet loss (`packet_loss`), and a list of results that detail, for each try, the IP address (`ip_address`) that was pinged and the round-trip time (`rtt`).

The fact that NAPALM does not return unstructured text but instead returns structured data in the form of dictionaries is a great benefit, as this means that we don't need to manually parse textual output in order to retrieve information. For example, you could use this functionality to check the connectivity of all your devices in intervals and report back whether a device fails to reach a host that it should be able to reach or whether the average round-trip time is significantly higher than what you would expect.

While NAPALM tries to offer all functions with all drivers, and the ping function has been chosen for this example because it is supported by all drivers, some functions are not. You can find a list of all the functions your specific driver supports in the official documentation. This list is currently available at `https://napalm.readthedocs.io/en/latest/support/index.html#getters-support-matrix`. We will examine more of these supported getters in the *Gathering facts about your device using NAPALM* recipe.

In this example, we have passed the username and password as variables in the script. While this is great for testing purposes, you can find a more secure way of passing on these variables in the *There's more...* section of the *Connecting to devices from different vendors using NAPALM* recipe.

Backing up your device configuration using NAPALM

With our vendor-agnostic functions, we can also carry out common backup tasks with ease. For instance, let's assume, before a big change, you want to back up all the configurations on your devices. NAPALM offers a very convenient way of retrieving not only the running configuration but also the starting and candidate configurations of your devices. In this recipe, we are going to explore how to use this ability to easily retrieve full configurations from a device to create backups.

Getting ready

Open your code editor and create a file called `backup_config.py`. Next, in your Terminal, navigate to the same directory that you just created the `backup_config.py` file in.

How to do it...

The following steps demonstrate how to connect to a device, read the different types of configurations, and then write them to a file:

1. Import the `napalm` module:

```
import napalm
```

2. From the `napalm` module, retrieve the appropriate driver for your device's vendor. Please refer to the *How it works...* section of the *Connecting to devices from different vendors using NAPALM* recipe for a list of possible drivers:

```
driver = napalm.get_network_driver("<insert driver
name>")
```

3. Specify the connection details of your device. Note that you can pass a non-standard port in the optional arguments. If your device connects on the standard port, you can omit this:

```
conn_details = {
        "hostname": "<insert hostname>",
        "username": "<insert username>",
        "password": "<insert password>",
        "optional_args": {
                "port": <insert port as integer>
        }
}
```

4. With the driver and connection details specified, we can create a new device:

```
device = driver(**conn_details)
```

5. Open a connection to the device:

```
device.open()
```

6. With our device connected, we can use the `get_config()` method to retrieve the different configurations:

```
config = device.get_config()
```

7. Retrieve the hostname from the connection details and save it to a variable. This is so that we can later use it as part of the filename for our backup:

```
host = conn_details['hostname']
```

8. The returned config is not a plain string but a dictionary where the keys specify the different configuration types; here, they are the starting, running, and candidate configurations. We can loop over all the keys that are present and create a backup of our config that includes the hostname:

```
for conf_type in config.keys():
        with open(f"{host}-{conf_type}.conf.bak", "w") as f:
                f.writelines(config[conf_type])
```

9. Finally, close the device:

```
device.close()
```

10. To run this script, go to your Terminal and execute it using the following command:

```
python3 backup.py
```

After the script has finished executing, you should have a new `.conf.bak` file for each of the configuration types in the directory.

How it works...

First, we establish a connection to our device by specifying the connection details and retrieving the correct driver for our vendor. Please refer to the *Connecting to devices from different vendors using NAPALM* recipe for a more detailed explanation of how to establish a connection to a device using NAPALM.

With our connection established, we then use the `get_config()` function to get a configuration. This function returns a dictionary where the keys describe the different types of configurations (for instance, running, starting, or candidate), and the corresponding values are the contents of the configuration itself. To create a backup of each of these different configurations, we loop over all of the keys in our config dictionary and open a new file. The name of this file always includes the host that we retrieved from the connection details along with the configuration type, and it ends in a `.conf.bak` file type. Opening those files after running the script will show the backed-up configuration.

In this example, we have passed the username and password as variables in the script. While this is great for testing purposes, you can find a more secure way of passing on these variables in the *There's more* section of the *Connecting to devices from different vendors using NAPALM* recipe.

Gathering facts about your network device using NAPALM

When dealing with a device, we are usually doing two different types of operations. For example, we are either setting information or we are gathering information. NAPALM offers us two different types of functions to carry out these types of actions. To set information, we have configuration functions, and to retrieve information, we have getter functions. We will explore configuration functions in the *Creating and applying a configuration template with jinja2 and NAPALM* recipe. We have already seen a getter, the `get_config()` function, in action in the *Backing up your device configuration using NAPALM* recipe. In this recipe, we will explore more of these functions so that we can retrieve more facts about the device we are connected to.

Getting ready

Open your code editor and create a file called `gather_facts.py`. Next, in your Terminal, navigate to the same directory that you just created the `gather_facts.py` file in.

How to do it...

The following steps demonstrate how to connect to a device and retrieve some facts about it using the getters that NAPALM provides:

1. Import the `napalm` module:

```
import napalm
```

2. From the `napalm` module, retrieve the approprtiate driver for your device's vendor. Please refer to the *How it works...* section of the *Connecting to devices from different vendors using NAPALM* recipe for a list of possible drivers:

```
driver = napalm.get_network_driver("<insert driver
name>")
```

3. Specify the connection details of your device. Note that you can pass a non-standard port in the optional arguments. If your device connects on the standard port, you can omit this:

```
conn_details = {
        "hostname": "<insert hostname>",
        "username": "<insert username>",
        "password": "<insert password>",
        "optional_args": {
                "port": <insert port as integer>
        }
}
```

4. With the driver and connection details specified, we can create a new device:

```
device = driver(**conn_details)
```

5. Open a connection to the device:

```
device.open()
```

6. With our device connected, we can now use the getter to retrieve some information. First, let's use the `get_facts()` function to retrieve general facts, such as the software version and the uptime:

```
facts = device.get_facts()
```

7. Next, we also want to retrieve the interfaces that are configured on this device:

```
interfaces = device.get_interfaces()
```

8. With our facts retrieved, we can now print them back to the user. NAPALM returns the dictionaries again. For the facts, we'll start by printing out the key-value pairs:

```
print("Key facts about your device:")
for fact, value in facts.items():
    print(f"-> {fact}: {value}")
```

9. For our interfaces, the `get_interfaces()` function returns a dictionary where the key is the name of the interface, and the value is another dictionary that contains facts such as the Mac address or description. We can use two nested `for` loops to print out this information to the user:

```
print("Facts about your devices interfaces")
for intf_name, details in interfaces.items():
    print(f"{intf_name}")
    for fact, value in details.items():
        print(f"=> {fact}: {value}")
```

10. Finally, close the connection to the device:

```
device.close()
```

11. To run this script, go to your Terminal and execute it using the following command:

```
python3 gather_facts.py
```

After the script has finished executing, you should have a list of key facts such as the OS version and the uptime of your device along with a list of information about each of the device's interfaces printed back to you.

How it works...

First, we establish a connection to our device by specifying the connection details and retrieving the correct driver for our vendor. Please refer to the *Connecting to devices from different vendors using NAPALM* recipe for a more detailed explanation of how to establish a connection to a device using NAPALM.

With our connection established, we can use the getters provided by NAPALM to retrieve the information. In this example, we are using the `get_facts()` and `get_interfaces()` functions. As with everything in NAPALM, which getters are available to your specific device depends on the driver, and you can verify the functionality in the support matrix. The support matrix is available in the official documentation. Currently, it can be found at `https://napalm.readthedocs.io/en/latest/support/index.html#getters-support-matrix`. Now that we have gathered our general facts and the interface details, one immediate benefit is visible. The data has already been parsed and converted from the plain text that NAPALM receives from the device itself into structured Python data that we can leverage. In this example, we go ahead and, first, use a simple loop to loop over the key-value pairs of our simple facts. The simple facts include, among other things, the following:

- `uptime`
- `vendor`
- `os_version`
- `serial number`
- `hostname`

Then, we also loop over all of our configured interfaces. The interfaces come as a nested dictionary where each interface's name is associated with a dictionary that contains, among other things, the details for that specific interface. These can include the following:

- `ip address`
- `enabled status`
- `active status`
- `mac address`
- `mtu`
- `description`

Using two nested `for` loops, we can parse this information and present it to our users in a consumable manner.

In this example, we have passed the username and password as variables in the script. While this is great for testing purposes, you can find a more secure way of passing on these variables in the *There's more...* section of the *Connecting to devices from different vendors using NAPALM* recipe.

Creating and applying a configuration template with jinja2 and NAPALM

A very powerful concept of NetDevOps is configuration templates. The idea is to have all your device's configurations as a template and then render out that template with some specific data for your device. This brings you one step closer to the idea of having your infrastructure stored as code. In this recipe, we will use the jinja2 templating language to create a configuration template that we can then apply to our device using NAPALM. Please refer to *Chapter 3, Building Configuration Templates Using Jinja2*, for a complete introduction to the jinja2 module and the jinja2 templating language. In this recipe, we are going to replicate the configuration template that we created in the *Configuring an access list using for loops in jinja2* recipe of *Chapter 3, Building Configuration Templates Using Jinja2*.

Getting ready

Open your code editor and create a file called `apply_template.py`. Next, in your Terminal, navigate to the same directory that you just created the `apply_template.py` file in.

Additionally, we'll need a jinja2 template. So, in the same directory as your Python file, create a directory called `templates`. Inside of this directory, create a file called `acl.conf.tpl`.

We'll be reusing the same template that we rendered to a file in the *Using for loops in jinja2 to configure an access list* recipe of *Chapter 3, Building Configuration Templates Using Jinja2*.

If you have not installed the jinja2 package from *Chapter 3, Building Configuration Templates Using Jinja2*, please go ahead and do so now. You can install the latest version of jinja2 using `python3 -m pip install jinja2`. At the time of writing, the current version is version `2.11.2`. You can install this specific version by using the `python3 -m pip install jinja2==2.11.2` command.

How to do it...

The following steps demonstrate how to render a configuration template and apply it to your device:

1. Open the `acl.conf.tpl` file. This is the template that is being rendered. First, we are going to specify the interface for our ACL and then loop over the two lists we have provided as an argument to create the necessary commands to allow or deny the host from the network:

```
interface {{ intf }}
ip access-group 1 in
{% for host in disallowed %}
access-list 1 deny host {{ host }}
{% endfor %}
{% for host in allowed %}
access-list 1 permit host {{ host }}
{% endfor %}
```

2. Continuing in our `apply_template.py` file, import the `napalm` module and the `jinja2` module:

```
import napalm
from jinja2 import Environment, FileSystemLoader
```

3. Next, we need to set up our `jinaj2` environment to render the template:

```
loader = FileSystemLoader("templates")
environment = Environment(loader=loader)
tpl = environment.get_template("acl.conf.tpl")
```

4. Now that we have our template, we need to define two lists that hold the IPs we want to deny and those we want to allow. Additionally, we are going to specify the interface that we want this ACL to be configured on in the following manner:

```
allowed = [
      "10.10.0.10",
      "10.10.0.11",
      "10.10.0.12"
]
disallowed = [
```

```
        "10.10.0.50",
        "10.10.0.62"
]
intf = "ethernet0"
```

5. With the information that is needed to render our template now in place, we can create a configuration from our template and store it in a variable:

```
out = tpl.render(allowed=allowed,
                 disallowed=disallowed,
                 intf=intf)
```

6. With our configuration rendered, we are ready to connect to our device. From the `napalm` module, retrieve the appropriate driver for your device's vendor. Please refer to the *How it works...* section of the *Connecting to devices from different vendors using NAPALM* recipe for a list of possible drivers:

```
driver = napalm.get_network_driver("<insert driver
name>")
```

7. Specify the connection details of your device. Note that you can pass a non-standard port in the optional arguments. If your device connects on the standard port, you can omit this:

```
conn_details = {
      "hostname": "<insert hostname>",
      "username": "<insert username>",
      "password": "<insert password>",
      "optional_args": {
            "port": <insert port as integer>
      }
}
```

8. With the driver and connection details specified, we can create a new device:

```
device = driver(**conn_details)
```

9. Open a connection to the device:

```
device.open()
```

10. With our device connected, we can now use the `load_merge_candidate()` function to load the rendered template as a candidate for merging:

```
device.load_merge_candidate(config=out)
```

11. The previous step has only loaded the config for merging. In order to apply it, we have to commit the configuration using the `commit_config()` function:

```
device.commit_config()
```

12. Finally, close the device connection:

```
device.close()
```

13. To run this script, go to your Terminal and execute it using the following command:

```
python3 apply_template.py
```

After the script has finished executing, your device should have the newly configured ACL present.

How it works...

First, we create a jinja2 template for our configuration. Once this template has been created, we can populate it with information from our Python script. For a more detailed explanation of how to render a configuration template, please refer to the *Using for loops in jinja2 to configure an access list* recipe of *Chapter 3, Building Configuration Templates Using Jinja2.*

With the config created, we can establish a connection to our device by specifying the connection details and retrieving the correct driver for our vendor. Please refer to the *Connecting to devices from different vendors using NAPALM* recipe for a more detailed explanation of how to establish a connection to a device using NAPALM.

Once the connection to the device has been established, we can use one of the setter functions, `load_merge_candidate()`, to load the configuration we just generated from the template as a candidate. By using the `load_merge_candidate()` function, we tell NAPALM to merge our existing configuration. We could also use the `load_replace_candidate()` function to simply replace the current configuration with our provided configuration. Once we have loaded this, the new configuration is set as a replacement candidate. This means that the config has not yet been applied and is not yet in action. To turn our replacement candidate into a running configuration, we have to use the `commit_config()` function to commit the configuration to the device.

In this example, we have passed the username and password as variables in the script. While this is great for testing purposes, you can find a more secure way of passing on these variables in the *There's more...* section of the *Connecting to devices from different vendors using NAPALM* recipe.

Rolling back configuration changes using NAPALM

Loading a configuration onto a device blindly, without verifying which lines actually change, can be a dangerous game, especially if the configuration that is being loaded has hundreds of lines or was generated programmatically. So, wouldn't it be nice if we could get a visual representation of all the changes that will be applied to our configuration and then decide whether we actually want to go ahead and apply them or discard the changes?

This is exactly what NAPALM allows you to do by using the `compare_config()` function together with `commit_config()` and `rollback()`.

Getting ready

Open your code editor and create a file called `rollback.py`. Next, in your Terminal, navigate to the same directory that you just created the `rollback.py` file in.

How to do it...

The following steps demonstrate how to connect to a device, view the differences between the new configuration and the old configuration, and then lets you choose whether you want to apply the configuration or not:

1. Import the `napalm` module:

```
import napalm
```

2. From the `napalm` module, retrieve the appropriate driver for your device's vendor. Please refer to the *How it works...* section of the *Connecting to devices from different vendors using NAPALM* recipe for a list of possible drivers:

```
driver = napalm.get_network_driver("<insert driver
name>")
```

3. Specify the connection details of your device. Note that you can pass a non-standard port in the optional arguments. If your device connects on the standard port, you can omit this:

```
conn_details = {
        "hostname": "<insert hostname>",
        "username": "<insert username>",
        "password": "<insert password>",
        "optional_args": {
                "port": <insert port as integer>
        }
}
```

4. With the driver and connection details specified, we can create a new device:

```
device = driver(**conn_details)
```

5. Open a connection to the device:

```
device.open()
```

6. With our device connected, first, we create a new configuration. In this example, we are just going to change the hostname. Each entry in the list represents one line of our configuration:

```
config_change = [
        "hostname test"
]
```

7. Since NAPALM requires a string as the replacement config, we join all the entries of our `config_change` list with newlines:

```
new_config "\n".join(config_change)
device.load_merge_candidate(config=new_config)
```

8. Then, we compare the differences between the currently running configuration and the configuration loaded for merging and display this information:

```
print(device.compare_config())
```

9. With the difference represented visually, we can ask the user to specify whether to either apply the configuration candidate or disregard it:

```
user_in = input("Continue? [y/n] ")
if user_in == "y":
        device.commit_config()
        print("Applied config to device")
else:
        device.rollback()
        print("Rolled back to previous config")
```

10. Finally, close the connection to the device:

```
device.close()
```

11. To run this script, go to your Terminal and execute it using the following command:

```
python3 rollback.py
```

How it works...

First, we establish a connection to our device by specifying the connection details and retrieving the correct driver for our vendor. Please refer to the *Connecting to devices from different vendors using NAPALM* recipe for a more detailed explanation of how to establish a connection to a device using NAPALM.

Once our connection has been established, first, we create a configuration string from a list of configuration changes. With this target configuration created, we can use the `load_merge_candidate()` function to load the configuration. After this loading is complete, we can use the `compare_config()` function to get a visual comparison of the changes. This is similar to how version control systems display the differences between two versions of a file (also known as a "diff"). A + sign in front of a line indicates that this is new, while a – sign indicates that this line was present in the old configuration and is no longer in the new configuration. With this visual representation, we can then ask the user for a decision. The user can either chose to apply the configuration by committing it or roll back to the previous version. As with most things in NAPALM, you have to verify whether the driver supports these kinds of comparisons. You can find the list of drivers supporting this operation in the official documentation. Currently, it is available at `https://napalm.readthedocs.io/en/latest/support/index.html#configuration-support-matrix`.

In this example, we have passed the username and password as variables in the script. While this is great for testing purposes, you can find a more secure way of passing on these variables in the *There's more...* section of the *Connecting to devices from different vendors using NAPALM* recipe.

Validating deployments using NAPALM

We have learned how NAPALM allows us to, very conveniently, issue commands, load and replace configurations, and gather facts. But what happens after you have carried out your changes? How do you verify that your device is still compliant with the state that you want to have? We'll take an even deeper look at automating the testing of your network devices in *Chapter 7, Automating Your Network Tests and Deployments with pyATSand Genie*. However, NAPALM also offers you a way to describe the desired state of your device in a human-readable format and make sure that the device actually complies.

The mechanism NAPALM uses to implement these are getters. Recall that these functions return well-formatted and structured Python data and have a unique function name. NAPALM leverages that by allowing you to describe what the output of this getter *should* be in a `.yaml` file and then runs the getter to find out what the device is currently running on.

Getting ready

Open your code editor and create a file called `validate_deployment.py`. Next, in your Terminal, navigate to the same directory that you just created the `validate_deployment.py` file in.

We'll also need a file called `compliance.yaml`. This needs to be in the same directory as your Python script.

How to do it...

To verify the compliance of a deployment based on a `.yaml` file describing the desired state, follow these steps:

1. Let's start by defining the desired state in our `compliance.yaml` file:

```
---
- get_facts:
    os_version: 4.17
- get_interfaces_ip:
    GigabitEthernet1:
```

```
        ipv4:
          10.10.10.1
```

2. Going back to our `validate_deployment.py`, we can now import the `napalm` module:

```
import napalm
```

3. From the `napalm` module, retrieve the appropriate driver for your device's vendor. Please refer to the *How it works...* section of the *Connecting to devices from different vendors using NAPALM* recipe for a list of possible drivers:

```
driver = napalm.get_network_driver("<insert driver
name>")
```

4. Specify the connection details of your device. Note that you can pass a non-standard port in the optional arguments. If your device connects on the standard port, you can omit this:

```
conn_details = {
      "hostname": "<insert hostname>",
      "username": "<insert username>",
      "password": "<insert password>",
      "optional_args": {
            "port": <insert port as integer>
      }
}
```

5. With the driver and connection details specified, we can create a new device:

```
device = driver(**conn_details)
```

6. Open a connection to the device:

```
device.open()
```

7. With our device connected, we can run our compliance report using the `compliance_report()` function:

```
out = device.compliance_report("compliance.yaml")
print(out)
```

8. Finally, close the connection to the device:

```
device.close()
```

9. To run this script, go to your Terminal and execute it using the following command:

```
python3 validate_deployment.py
```

How it works...

First, we define the desired state of our device in the YAML file. As you can see, we are using the names of the getters as the keys and then specifying the values we want. In our example, we specify that our device should have an OS version of 4.17 and that GigabitEthernet1 should have an IPv4 address of `10.10.10.1`. The IPv4 validation example shows that we can also check nested data. Our interface for example has nested properties such as, in this example, the IPv4 address of one of the interfaces.

With our desired state described, we can establish a connection to our device by specifying the connection details and retrieving the correct driver for our vendor. Please refer to the *Connecting to devices from different vendors using NAPALM* recipe for a more detailed explanation of how to establish a connection to a device using NAPALM.

Once the connection has been established, we run the compliance report. This will invoke all the getters that are specified in our `compliance.yaml` file and compare the values. What we get back is a dictionary that defines all the parts of this device that are either in or out of compliance. We can see the value that was expected along with the value that we got. This can be a powerful tool to verify, after a big change, that all the changes have also been applied properly or that things that you have not touched, such as BGP routes, for example, are still configured to how you had them before.

In this example, we have passed the username and password as variables in the script. While this is great for testing purposes, you can find a more secure way of passing on these variables in the *There's more...* section of the *Connecting to devices from different vendors using NAPALM* recipe.

7
Automating Your Network Tests and Deployments with pyATS and Genie

In previous chapters, we have seen how we can use Python and open source packages to automatically retrieve and change a configuration on one or more network devices. When we think about the normal process we go through when manually updating infrastructure configuration, the next step would be to test that the configuration change we just initiated didn't have any unwanted side effects. How could we automate this step? With the packages we previously discussed, this would be quite cumbersome since we'd have to manually retrieve the state of the device before and after our change and then do the differentiation between the two versions ourselves.

This is where pyATS and Genie come into play. Originally developed as an internal testing framework at Cisco, the framework was open sourced and now supports the deployment and testing of changes for network devices. Build-agnostic by design, the framework supports plugins that extend its compatibility and functionality beyond Cisco devices. With its broad scope, pyATS and Genie allow you to tackle all tasks of automatically rolling out changes to your infrastructure, from planning to deploying and testing.

In this chapter, we will explore the capabilities of pyATS and Genie to both deploy and then test the changes you have made to your infrastructure. While pyATS and Genie are technically two libraries—pyATS being responsible for device connectivity, test definition, and reporting, and Genie being responsible for high-level capabilities such as parsing **command-line interface** (**CLI**) output to structured Python data or defining test cases—we will refer to them both as pyATS for the remainder of this chapter.

Specifically, we will cover the following topics:

- Revisiting the concept of testing
- Creating a pyATS testbed file
- Connecting to your device and issuing commands using pyATS
- Retrieving your device's current state using pyATS
- Using Genie `Conf` objects to create a portable configuration script
- Comparing your device's current state to a previously learned state

Technical requirements

For this section and the remainder of the book, you'll need an installation of Python. Specifically, you'll need a Python interpreter of version 3.6.1 or higher. This book makes use of language constructs of Python 3 and thus is incompatible with Python 2.x. Additionally, you'll need to install the pyATS and Genie packages. You can install the newest version of pyATS and Genie using `python3 -m pip install pyats`. At the time of this writing, the current version is version 21.1.

You will also need a code editor. Popular choices include Microsoft **Visual Studio Code** (**VS Code**) or Notepad++. Additionally, you'll need a device (virtual or physical) that you can log in to via **Secure Shell** (**SSH**).

Please be aware that pyATS and Genie only work on Linux and Linux-like environments such as Ubuntu, CentOS, or Fedora Linux distributions, as well as the family of macOS operating systems. If you are using Windows, you can run pyATS either in a Linux **virtual machine** (**VM**) or by using the **Windows Subsystem for Linux** (**WSL**).

You can view this chapter's code in action here: `https://bit.ly/3k4TabN`

Revisiting the concept of testing

So, why do we need testing and, more specifically, why would we want to automate it? A lot of the ideas that have been proposed for network automation have their roots in software engineering, so it is worth having a brief look at how modern software is developed and deployed on a continuous basis. When developing software, most engineers go through four basic phases, outlined as follows:

1. The **development stage** is where the code is first designed and then written.
2. The **testing stage** is where the code that has been written is tested extensively.
3. The **deployment stage** is where the code that has been developed and tested is deployed to the servers.
4. The **monitoring stage** is where data such as access times, occurring errors, and load on the system of the new change is closely monitored. Any insights that have been identified in the monitoring stage (such as the need to optimize a certain portion of the code) are fed back into the next development stage, leading to an infinite cycle of developing, testing, deploying, and monitoring your software.

While this is a simplified model, it gives us an idea of the different steps involved in developing, deploying, and maintaining software. As with infrastructure, every change to a software system is inherently risky since we are changing something we knew worked and replacing it with code that might not work. In order to make smaller and thus usually safer changes, we need to make *Steps 2-4* as effortless as possible. If it takes our operations team a day to take a change by a developer and deploy it to the production systems, the developers will be hesitant to ask for changes to be applied frequently, thus leading to bigger and bigger changes that are inherently riskier. This is where automation comes into play. By automating the testing, deployment, and monitoring stage, we can deploy smaller changes to the code base much more often and react faster to issues. Instead of releasing once a month or once a week, some software companies these days measure their deployment speed in changes per minute.

Going back to infrastructure engineering, we can identify a similar cycle, outlined as follows:

1. The **design phase** is where changes to our infrastructure are designed and then implemented in the form of a change of configuration.
2. The **testing phase** is where the changes are usually applied in a lab environment to troubleshoot and identify any potential issues.
3. The **deployment phase** is where, during a maintenance window, the different devices are being updated with the new configuration.
4. The **monitoring phase** is where issues that may have occurred are monitored.

Similar to how automation can be used to speed up the change cycle in software products, we can use it to speed up the change cycle in our infrastructure. Speeding up this change cycle can lead to smaller changes that, similar to smaller software changes, are less risky. And similar to how, with a faster software deployment cycle, we can fix issues in our software faster, we can leverage a sped-up deployment cycle of our infrastructure to resolve issues faster.

While the cycles of software change and infrastructure change are quite similar, there is one key difference: it is much easier to test software. With a software project that has been changed, you can just spin up a new server, deploy the software—including all its dependencies, such as databases—to that new server, and then run a bunch of tests against the newly updated instance. Once the testing is done, the server can be shut down. With infrastructure, this is not as easy. Usually, you'll have a lab environment, either virtual or physical, that can be used to test changes before applying them into production.

This means that when we are doing testing in an infrastructure environment, we will often do testing before the changes are applied by testing on our lab environment as well as after the changes have been applied to the production environment, to verify that no unwanted side effects have been occurring. The framework presented in this chapter, pyATS, can be used to perform all four stages of the preceding cycle of making a change to the infrastructure. In this chapter, we will put our focus on how to use pyATS for designing, deploying, and testing your infrastructure changes.

Creating a pyATS testbed file

When connecting to devices, we always need a way of specifying the connection details. In previous chapters, we have provided these details either in the script or asked the user to provide the connection details upon execution. With pyATS, there is a different concept. pyATS relies on a text file, written in **YAML Ain't Markup Language** (**YAML**) and commonly referred to as a *testbed*, that specifies the different attributes such as device type, hostname, and username/password for our device.

In this recipe, you'll see how to create such a testbed file and how to use the interactive shell that comes with every installation of pyATS to test the connectivity to the devices in your testbed.

Getting ready

Open your code editor and start by creating a file called `testbed.yaml`. Next, navigate in your terminal to the same directory in which you just created the `testbed.yaml` file.

How to do it...

Follow these steps to define a testbed for your devices to connect to:

1. Open your testbed.yaml file and define a devices list and a hostname for your device—in this case, device-1:

```
devices:
  device-1:
```

2. Next, define the device type (in this example, router) and the operating system (in this example, iosxe) running on the device, as well as the connection credentials:

```
    type: router
    os: iosxe
    credentials:
      default:
        username: <insert your username>
        password: <insert your password>
```

3. Next, we need to define the connection details to our management interface. In this example, we are going to connect to the device via SSH. If you want to use an **Internet Protocol** (**IP**) address instead of a host, use ip: <insert your ip here> instead of the host part shown here:

```
    connections:
      mgmt:
        protocol: ssh
        host: <insert your host here>
        port: <insert your port here>
```

4. With the testbed file created, we can go ahead and use the pyATS shell to connect to our device. Open your console in the same directory as the testbed file and type the following command:

```
pyats shell -testbed-file testbed.yaml
```

5. After the shell has loaded and you have an interactive Python shell, type the following command to connect your device and open a device shell within the Python session:

```
testbed.devices['device-1'].connect()
```

How it works...

In our YAML file, we are defining properties that are needed for pyATS to be able to connect. The first directive, `devices`, specifies that we are about to define a list of devices. Underneath this list of devices, we then give each of these devices a name—in our example, `device-1`. You can use any name you want as long as it is unique, but most people prefer to use the hostname of their device here. Next, we define the `type` and `os` so that pyATS knows how to connect to this device. Out of the box, pyATS ships with support for connecting to all Cisco **Internetwork Operating System** (**IOS**) variants, as well as the Nexus **Operating System** (**OS**) family.

Finally, we specify the authentication details such as username and password, as well as the connection details to our management interface. pyATS supports SSH, **RESTCONF** (which stands for **REpresentational State Transfer** (**REST**) **configuration**), and **NETCONF** (which stands for **network configuration**), but we will stick to SSH for this chapter.

With our device specified within the testbed, we can use the interactive shell provided by pyATS to verify that we have specified everything correctly and can connect to our device. When starting the shell, we specify the name of the testbed file, and pyATS automatically imports the required modules of the framework that parse our testbed file and provides us with a `testbed` object. This object has a dict-like property called `devices` that we can call to—based on the name we specified in the testbed file—retrieve our device and connect to it using the `connect()` method. This will open a shell to the device that we can use to issue a device command such as `show ip interface brief`.

Connecting to your device and issuing commands using pyATS

A basic functionality we have seen across all the different packages covered so far is the ability to issue commands against a device. In the *Retrieving command outputs as structured Python data using Netmiko and Genie* recipe in *Chapter 4, Configuring Network Devices Using Netmiko*, we have already seen that we can use Genie, the package that is part of the pyATS ecosystem, to retrieve the output of a command not as text but as structured Python data.

In this recipe, we are going to use pyATS to connect to a device, issue a command, and retrieve the output as structured data.

Getting ready

Open your code editor and start by creating a file called `testbed.yaml`, as well as a file called `connect.py`. Next, navigate in your terminal to the same directory in which you just created the `testbed.yaml` file.

How to do it...

Follow these steps to connect to a device and issue a command using pyATS:

1. Open your `testbed.yaml` file and define a `devices` list and a hostname for your device—in this case, `device-1`:

```
devices:
  device-1:
```

2. Next, define the device type (in this example, `router`) and the operating system (in this example, `iosxe`) running on the device, as well as the connection credentials:

```
    type: router
    os: iosxe
    credentials:
      default:
        username: <insert your username>
        password: <insert your password>
```

3. Next, we need to define the connection details to our management interface. In this example, we are going to connect to the device via SSH. If you want to use an IP address instead of a host, use `ip: <insert your ip here>` instead of the `host` part shown here:

```
    connections:
      mgmt:
        protocol: ssh
          host: <insert your host here>
          port: <insert your port here>
```

4. With the testbed file created, we can go ahead and open the `connect.py` file. We first need to import the required pyATS library to load the testbed file:

```
from pyats.topology import loader
```

5. Next, load the testbed file and retrieve the device by its name (in this example, `device-1`), and issue a command:

```
testbed = loader.load('testbed.yaml')
device = testbed.devices['device-1']
device.connect()
```

6. With the device connection established, we can issue our command and retrieve the output as unstructured data:

```
out = device.execute('show ip interface brief')
print(out)
```

How it works...

We first create our testbed file. This file specifies the details of all the devices that we want to be able to connect to and makes them available to pyATS. For a more detailed explanation of how this testbed file is constructed, please refer to the *How it works...* section of the *Creating a pyATS testbed file* recipe.

With our testbed file created, we can then import the loader from the pyATS `topology` module. This loader parses the testbed file into a topology and makes the devices available to our Python script. Using the name we have specified in the testbed file, we can then retrieve the device we want to connect to, connect to it, and execute a command. The output of that command is retrieved in text form and Genie uses its parsers to convert it from text into structured Python data, in the form of a dictionary that is then printed back to the user.

Retrieving your device's current state using pyATS

With pyATS, we have the ability to retrieve the output of a network device command as structured Python data. With this structured data, we can get a full picture of what the configuration of our device looks like. Now, could we use this ability of pyATS to *learn* the state of a network device? Using this learned state, we could take a snapshot of our device after applying a configuration change, and then, the next time we want to apply a change to this device, we can verify that no one has manually applied changes to the configuration by comparing the snapshot we took before with the current state of the device. This is exactly the scenario that we are going to cover in the *Comparing your device's current state to a previously learned state* recipe, and in this recipe we are going to cover the first part of this, learning the state of a device using pyATS.

Getting ready

Open your code editor and start by creating a file called `testbed.yaml`, as well as a file called `learn.py`. Next, navigate in your terminal to the same directory in which you just created the `testbed.yaml` file.

How to do it...

Follow these steps to *learn* the current state of a network device automatically using pyATS:

1. Open your `testbed.yaml` file and define a `devices` list and a hostname for your device—in this case, `device-1`:

```
devices:
    device-1:
```

2. Next, define the device type (in this example, `router`) and the operating system (in this example, `iosxe`) running on the device, as well as the connection credentials:

```
        type: router
        os: iosxe
        credentials:
            default:
                username: <insert your username>
                password: <insert your password>
```

3. Next, we need to define the connection details to our management interface. In this example, we are going to connect to the device via SSH. If you want to use an IP address instead of a host, use `ip: <insert your ip here>` instead of the `host` part shown here:

```
        connections:
            mgmt:
                protocol: ssh
                host: <insert your host here>
                port: <insert your port here>
```

4. With the testbed file created, we can go ahead and open the `learn.py` file. We first need to import the required pyATS library to load the testbed file. We will also import the built-in **JavaScript Object Notation** (**JSON**) library to save the output of our learning to a file:

```
from genie.testbed import load
import json
```

5. Next, load the testbed file and retrieve the device by its name (in this example, `device-1`), and issue a command:

```
testbed = load('testbed.yaml')
device = testbed.devices['device-1']
device.connect()
```

6. With our device connected, we can initiate the learning. In this example, we are going to learn all available device features:

```
output = device.learn('all')
```

7. Lastly, we'll have to write the data stored in our output variable to a file for us to later retrieve it:

```
with open("backup.json, "w") as fh:
      json.dump(output.to_dict(), fh, indent=2)
```

How it works...

We first create our testbed file. This file specifies the details of all the devices that we want to be able to connect to and makes them available to pyATS. For a more detailed explanation of how this testbed file is constructed, please refer to the *How it works...* section of the *Creating a pyATS testbed file* recipe.

With our testbed file created, we can then import the loader from the pyATS `topology` module. This loader parses the testbed file into a topology and makes the devices available to our Python script. Using the name we have specified in the testbed file, we can then retrieve the device we want to connect to and connect to it. We can then use the `learn()` function to learn a feature. In this example, we used `all` to indicate to pyATS that we want to learn all the available features on this platform. You can also only learn certain parts of the configuration—such as `interface`, `bgp`, or `ospf`—by specifying the name. A full list of available models can be accessed using this link: `https://pubhub.devnetcloud.com/media/genie-feature-browser/docs/#/models`.

You might be wondering why we can't just use the `execute()` function to issue the `show` commands ourselves and then write the parsed output to a JSON file for later diffing. First off, the `learn()` function, depending on which model you are learning, will invoke multiple `show` commands but, more importantly, the `learn()` function also ensures a *consistent* set of keys. This means that we can write one script that works across different devices. Once we have retrieved this output, we can then go ahead and save it to a file using the JSON module.

Using Genie Conf objects to create a portable configuration script

So far, we have always used the Genie parser to go from unstructured textual output to structured data in the form of a Python dictionary, but Genie can also do it the other way around! Instead of specifying a configuration command and issuing this command to the network device we can, using pure Python objects, define our changed configuration in the form of Python objects. Genie then goes ahead and, based on the Python objects, generates the required configuration files and issues them to the device. This means that we can write consistent device configurations and let Genie handle the work of converting this into *actual* configuration instructions.

In this recipe, we are going to configure an interface and generate the required configuration file using pyATS and Genie.

Getting ready

Open your code editor and start by creating a file called `testbed.yaml`, as well as a file called `change.py`. Next, navigate in your terminal to the same directory in which you just created the `testbed.yaml` file.

How to do it...

Follow these steps to change the configuration of an interface using pyATS:

1. Open your `testbed.yaml` file and define a `devices` list and a hostname for your device—in this case, `device-1`:

```
devices:
    device-1:
```

2. Next, define the device type (in this example, `router`) and the operating system (in this example, `iosxe`) running on the device, as well as the connection credentials:

```
type: router
os: iosxe
credentials:
    default:
        username: <insert your username>
        password: <insert your password>
```

3. Next, we need to define the connection details to our management interface. In this example, we are going to connect to the device via SSH. If you want to use an IP address instead of a host, use `ip: <insert your ip here>` instead of the `host` part shown here:

```
connections:
    mgmt:
        protocol: ssh
        host: <insert your host here>
        port: <insert your port here>
```

4. With the testbed file created, we can go ahead and open the `change.py` file. Here, we are going to import the required libraries to connect to our device based on the testbed file and also the required models to change our interface:

```
from pyats.topology import loader
from genie.conf.base import Interface
```

5. Next, we can load our device and connect to it:

```
testbed = loader.load('testbed.yaml')
device = testbed.devices['device-1']
device.connect()
```

6. Now, we can use the previously imported `Interface` object to create a new interface configuration:

```
intf = Interface(device=device, name='GigabitEthernet2')
```

7. With this `Interface` object, we can now specify the properties we want:

```
intf.ipv4 = '10.10.10.2'
intf.ipv4.netmask = '255.255.255.0'
intf.shutdown = False
```

8. Finally, we can apply the configuration to our interface by issuing a `build_config()` command:

```
intf.build_config()
```

How it works...

We first create our testbed file. This file specifies the details of all devices that we want to be able to connect to and makes them available to pyATS. For a more detailed explanation of how this testbed file is constructed, please refer to the *How it works...* section of the *Creating a pyATS testbed file* recipe.

With our device created and connected, we can go ahead and use the `Interface` object imported at the beginning to specify our interface configuration. On this object, we specify things such as the IP Address and netmask we would like to have on this interface. We can then let Genie generate the required configuration and automatically apply this configuration to our device using the `build_config()` function.

There's more...

You can also have a look at what the generated configuration looks like without actually applying the configuration right away. To do so, use the `apply=False` argument for the `build_config()` function:

```
print(intf.build_config(apply=False)
```

Running the preceding command will give us the following configuration:

```
interface GigabitEthernet2
 ip address 10.10.10.2 255.255.255.0
 no shutdown
 exit
```

You can also do the reverse, and generate the configuration necessary to *undo* the changes you have specified. To do so, instead of using the `build_config()` function, you can use the `build_unconfig()` function:

```
print(intf.build_unconfig(apply=False))
```

Running the preceding command will return to us the following commands, which would undo the previous configuration:

```
default interface GigabitEthernet2
interface GigabitEthernet2
shutdown
```

If we had left out the `apply=False` flag, the configuration would have actually been unapplied on our device.

Comparing your device's current state to a previously learned state

In the *Retrieving your device's current state using pyATS* recipe, we learned how to retrieve the state of a device, and in the previous recipe, *Using Genie Conf objects to create a portable configuration script*, we have seen how to apply changes to our device. This means that we can now not only save the state of a device but also change it. This recipe will cover the final piece of the puzzle and show how to compare our current device state to a previously saved state and print out the difference in the configuration.

Getting ready

Open your code editor and start by creating a file called `testbed.yaml`, as well as a file called `compare.py`. Next, navigate in your terminal to the same directory in which you just created the `testbed.yaml` file.

You'll also need to first create a snapshot of your device's features (see the *Retrieving your device's current state using pyATS* recipe) before applying some changes, either manually or automatically, to your device. For an introduction to adding changes automatically, see the *Using Genie Conf objects to create a portable configuration script* recipe.

How to do it...

Follow these steps to get the differences between a previously saved state and the current state of your device:

1. Open your `testbed.yaml` file and define a `devices` list and a hostname for your device—in this case, `device-1`:

```
devices:
    device-1:
```

2. Next, define the device type (in this example, `router`) and the operating system (in this example, `iosxe`) running on the device, as well as the connection credentials:

```
        type: router
        os: iosxe
        credentials:
            default:
                username: <insert your username>
                password: <insert your password>
```

3. Next, we need to define the connection details to our management interface. In this example, we are going to connect to the device via SSH. If you want to use an IP address instead of a host, use `ip: <insert your ip here>` instead of the `host` part shown here:

```
        connections:
            mgmt:
                protocol: ssh
                host: <insert your host here>
                port: <insert your port here>
```

4. With the testbed file created, we can go ahead and open the `compare.py` file. In it, we are going to first import the required libraries:

```
from genie.testbed import load
from genie.utils.diff import Diff
import json
```

5. Next, load the device from the testbed and connect to it:

```
testbed = load('testbed.yaml')
device = testbed.devices['device-1']
device.connect()
```

6. With the device loaded, we can learn the *current* state of this device:

```
current = device.learn('all')
```

7. After pyATS has learned the current state of the device, it's time to load the previously saved snapshot. In this example, we are assuming that the snapshot was saved to a file called `backup.txt`:

```
with open('backup.txt') as fh:
    snapshot = json.load(fh)
```

8. Finally, we can generate the difference between the previous state, retrieved from the loaded snapshot, and the current state of our device and print the difference back to the user:

```
diff = Diff(snapshot, current.to_dict())
diff.findDiff()
print(diff)
```

How it works...

We first create our testbed file. This file specifies the details of all devices that we want to be able to connect to and makes them available to pyATS. For a more detailed explanation of how this testbed file is constructed, please refer to the *How it works...* section of the *Creating a pyATS testbed file* recipe.

With our device created and connected, we can go ahead and retrieve the current state of the device by using the `learn()` function. For a detailed explanation of how the `learn()` function works, please refer to the *Retrieving your device's current state using pyATS* recipe. With our current state learned, we can then load the previously saved state from our JSON file and use the `diff` function from Genie to generate the difference between the two snapshots.

8
Configuring Devices Using RESTCONF and requests

In *Chapter 5, Model-Driven Programmability with NETCONF and ncclient*, we already saw how you can use **Yet Another Next Generation** (**YANG**) modules to model a network device's configuration and then use the **network configuration** (**NETCONF**) protocol to either retrieve or change the configuration of your device. The aim of NETCONF was to provide a unified interface for these operations. However, one major drawback of NETCONF is its rather unmodern reliance on **Extensible Markup Language** (**XML**) as the language of choice for describing the data as well as the operations you want to carry out. To solve this, the **Internet Engineering Task Force** (**IETF**) came up with the concept of **REpresentational State Transfer** (**REST**) **configuration** (**RESTCONF**). Built to use the same concepts as NETCONF, RESTCONF allows you to use the **HyperText Transfer Protocol** (**HTTP**) to interact with your device instead of the **Remote Procedure Calls** (**RPCs**) used by NETCONF. And, instead of forcing the user to use XML, RESTCONF lets us describe the data in either XML or **JavaScript Object Notation** (**JSON**).

In this chapter, we will learn how to use RESTCONF to carry out basic workflows around your devices. Specifically, we will look at the following topics:

- Revisiting HTTP's request-response model and RESTCONF principles
- Making HTTP requests using the `requests` module in Python

- Retrieving all interfaces of a device using RESTCONF and `requests`
- Creating a **Virtual Local-Area Network** (**VLAN**) using RESTCONF and `requests`
- Updating a VLAN using RESTCONF and `requests`
- Deleting a VLAN using RESTCONF and `requests`

Technical requirements

It is recommended that you have a basic understanding of how YANG and NETCONF work before diving into this chapter. For an introduction to both topics, please refer to the *Revisiting NETCONF and YANG modules* recipe in *Chapter 5, Model-Driven Programmability with NETCONF and ncclient.*

Since RESTCONF is built on top of HTTP requests, we will need the ability to make HTTP requests with Python. While we could implement the HTTP protocol ourselves or use one of the low-level modules that come with the Python standard library to perform these requests, the Python community has developed a package called `requests` that makes dealing with HTTP requests and responses much easier. Under the slogan *HTTP for Humans*™, this has abstracted most of the difficult tasks of dealing with HTTP on top of the standard library and has become the de facto standard for consuming HTTP-based resources from Python. We'll be using `requests` for this chapter as well as for *Chapter 9, Consuming Controllers and High-Level Networking APIs with requests.*

For this section and the remainder of the book, you'll need an installation of Python. Specifically, you'll need a Python interpreter of version 3.6.1 or higher. This book makes use of language constructs of Python 3 and thus is incompatible with Python 2.x. Additionally, you'll need to install the `requests` package. You can install the newest version of `requests` using `python3 -m pip install requests`. At the time of this writing, the current version is version 2.25.1. You can install this specific version of the `requests` module by using `python3 -m pip install requests==2.25.1`.

You will also need a code editor. Popular choices include Microsoft **Visual Studio Code** (**VS Code**) or Notepad++. Additionally, you will need a device (virtual or physical) that you can log in with via **Secure Shell** (**SSH**) and that has RESTCONF enabled. Please consult the manual of your device to find configuration instructions on how to enable the RESTCONF feature.

You can view this chapter's code in action here: `https://bit.ly/3yHLjqC`

Revisiting HTTP's request-response model and RESTCONF principles

When Tim Berners-Lee, a scientist at the **European Organization for Nuclear Research** (**CERN**) in Geneva, published the first drafts for HTTP, the protocol that would become the cornerstone of the **World Wide Web** (**WWW**), he probably did not expect the way his idea would develop. Started as a way for scientists to exchange their findings and collaborate on particle physics, life today could not be envisioned without the web.

In this section, we are going to revisit the basics of how HTTP works and have a specific look at how RESTCONF uses the ideas and building blocks defined by HTTP to model the configuration state and changes thereof of a network device. If you are already very familiar with the HTTP protocol itself, you may want to skip the next section and jump straight to the *How RESTCONF builds on top of HTTP* subsection of this recipe.

How does HTTP work?

HTTP is a plain text request-response protocol. A client sends an HTTP request to a server and receives an HTTP response back from the server.

An HTTP request always has three components, outlined here:

- A **request line** that specifies the **request method** and the **resource** that is accessed, identified by a resource **identifier** (**ID**)
- **Request header** fields that specify meta-information that the server should follow
- An optional **request body** that is separated from the request header fields by a newline

A request method is a specific keyword that describes the kind of operation we want the server to carry out when receiving and processing an HTTP request. The most important request methods, and the ones that will be relevant for our usage in RESTCONF and when dealing with REST **Application Programming Interfaces** (**APIs**), are outlined here:

- `GET` specifies that we are requesting to read information. This could be, for example, the retrieval of a device state.
- `POST` specifies that we want to submit the data enclosed in our request body to the specified resource. Data submitted via the `POST` method can be used to either create or update a resource.
- `PUT` specifies that we want to replace the resource identified with that specified in the request body.

- `PATCH` specifies that we want to overwrite or amend the resources identified by the resource ID with the data specified in the request body.
- `DELETE` specifies that we want to delete the resource specified by the resource ID.

The header fields of our HTTP request are used to specify meta-information that we want to give to our server when processing the request. An **HTTP header** is a key-value pair that is sent by the client to the server. While there are standard header fields as well as common non-standard fields, we can pass any information we want as part of a header. We just need a server that can interpret the information properly. The most common types of information passed inside of a header are outlined here:

- `Content-Type`: This header field specifies the type of data (such as JSON or XML) that is being sent as part of the request body. A `POST` request that contains JSON-encoded data in it would set the `Content-Type: application/json` header. This line tells the server that the body should be interpreted as JSON-formatted data.
- `Accept`: This header field specifies the way we would like to receive content back from the server. When we request information from our device and want to retrieve this information formatted as JSON, we would set an `Accept: application/json` header in our request header.
- `Authorization`: This header field is used to provide authentication credentials such as username/password or API tokens with our request. As an example, if the web server we are interacting with supports **basic authentication** (**HTTP Basic Auth**), which is authentication with a username/password combination, we would send a header like this: `Authorization: Basic cm9vdDpwYXNzd29yZA==`. The `Basic` keyword describes the type of authorization used, and the string that follows is a Base64-encoded representation of `<username>:<password>`. In this case, our username is `root` and the password is `password`, so the string that is being Base64-encoded is `root:password`. For RESTCONF, we will stick with this HTTP Basic Auth type. In the *Authenticating web requests* recipe of *Chapter 9, Consuming Controllers and High-Level Networking APIs with requests*, we will revisit some other concepts of authenticating web requests.
- `Content-Length`: This specifies the size of our request body in octets.
- `Host`: This specifies the host (and, optionally, the port) that the web server is listening on.

An authenticated HTTP `POST` request to the `/devices` resource, containing some JSON-encoded information that is being sent to a server running on port `8080` of host `data.com` that requests JSON data in response, would thus look like this:

```
POST /devices HTTP/1.1
Host: data.com:8080
Content-Type: application/json
Content-Length: 63
Accept: application/json
Authorization: Basic cm9vdDpwYXNzd29yZA==

{
    "name": "my device",
    "mac": "01-23-45-67-89-AB-CD-EF"
}
```

Our server will then come back and return to us an **HTTP response**. This response will either contain the information we have requested from the server in our HTTP request or an error. An HTTP response always has the following components:

- A **status line** with the protocol version and status code that is sent by the server to indicate whether a request was successful or not. An example for a status line could be `HTTP/1.1 200 OK`, which specifies that the HTTP protocol version is version `1.1`, the status code is `200`, and the reason phrase is `OK`.
- **Response header fields** that, as with request header fields, allow the server to pass meta-information such as the type of content being sent to the client. A server sending JSON-encoded information back to the client would, for example, specify this to the client by sending a `Content-Type: application/json; charset=utf-8` header that indicates to the client that the message body should be interpreted as JSON-encoded information and be encoded using `utf-8`.
- A (possibly empty) **message body** that contains the information we have requested from the server and that is separated from the header fields by a blank line.

An HTTP response to the HTTP request we have seen before could thus look like this:

```
HTTP/1.1 200 OK
Content-Type: application/json; charset=utf-8

{
```

```
    "id": 25,
    "name": "my device",
    "mac": "01-23-45-67-89-AB-CD-EF"
}
```

One of the most important concepts when dealing with HTTP requests, be it when using them for RESTCONF or when interacting with REST APIs, are the status codes. These codes can be used by the server developers to specify whether a request should be interpreted as a success or a failure and what kind of response was sent. Status codes have a possible range from `100` to `599` and are accompanied by a reason phrase. The reason phrase can be set freely by the application developer, but most status codes have reason phrases, such as `OK` for a response with status code `200`, which has become the norm.

Status codes come in five broad categories, outlined as follows:

- `1XX` for informational responses.
- `2XX` for successful responses.
- `3XX` for responses that redirect.
- `4XX` for responses that indicate a client error—for example, a missing authentication or a wrongly formatted request body that could not be understood by the server.
- `5XX` for responses that indicate a server-side error. This means that the request was formatted and understood correctly but the server encountered an internal error while processing the request.

The most important status codes that we will encounter when dealing with RESTCONF and REST APIs are listed next.

For the `2XX` range:

- `200 - OK`: A successful HTTP response. This is usually sent when doing `GET` requests to retrieve information but also when using `PATCH`/`PUT` requests to replace or update existing information.
- `201 - Created`: A resource was successfully created by the request that was sent. This response is mainly sent when sending information to a server via a `POST` request.
- `204 - No Content`: A response with this status code is returned when the request was successful but no data is returned. This code is mainly encountered when dealing with `DELETE` requests.

For the `3XX` range:

- `301 - Moved permanently`: This indicates that the request should be redirected.

For the `4XX` range:

- `400 - Bad request`: The server was unable to understand the request that was sent. This is usually due to a wrongly formatted request body or missing data in the request body.
- `401 - Unauthorized`: The server was expecting authorization information—for example, an `Authorization` header with a username/password combination according to the HTTP Basic Auth method, but the request did not supply that information.
- `403 - Forbidden`: The request included required authorization information, but the credentials provided do not have the required permissions. An account that only has read-only access to resources might encounter this error when sending an HTTP `DELETE` request with a valid `Authorization` header. The server would successfully authenticate the user, enabling them to realize that they do not have the required permissions to carry out a `DELETE` operation and thus the request is `Forbidden`, as outlined in the reason phrase.
- `404 - Not Found`: The requested resource could not be found on the server.
- `405 - Method Not Allowed`: The request used a method that is not allowed for this specific resource. Maybe your server only allows `GET` requests to the `/devices` resource and you are trying to send a `POST` request. In that case, this response code would be returned.

For the `5XX` range:

- `500 - Internal Server error`: The server has encountered some sort of error. A response with this error code means that the request could not be carried out correctly due to an issue on the server side.
- `502 - Bad Gateway`: The server had issues reaching another upstream server.
- `503 - Service Unavailable`: The server currently cannot handle the request—for example, because the web service you are interacting with is currently rolling out some updates.

You'll see that we will check these status codes frequently when dealing with RESTCONF or REST APIs to verify that our requests were successful. Here is a summary of an HTTP request-response process:

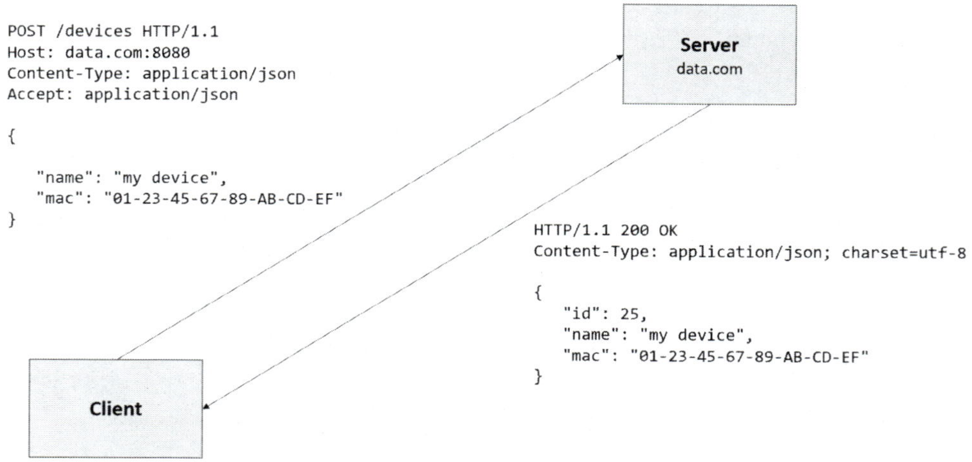

Figure 8.1 – Schematic summary of an HTTP request-response process

With the underlying HTTP protocol taken care of, let's see how RESTCONF builds on top of HTTP to provide a REST-like way of interacting with the configuration of a network device.

How RESTCONF builds on top of HTTP

So, how does RESTCONF use the concepts of HTTP, and where does it come from? With the rise of REST APIs, an API standard built on top of the HTTP protocol, the ask for a way to automate the change and retrieval of configuration data from network devices in a REST-like way became more and more prominent.

The idea behind RESTCONF was to build a REST-like interface that can be used similarly to NETCONF, but instead of using RCPs and XML, we use HTTP requests and either XML or JSON to format the data we are retrieving or submitting to the device.

We can still use the YANG modules to specify the data we want to retrieve or change and then use HTTP requests and JSON/XML to format and send the data to our device. Each operation in NETCONF is mapped to an HTTP request method, just as REST APIs map the **Create**, **Read**, **Update**, and **Delete** (**CRUD**) operations of a database system to HTTP methods, as specified in the following list:

- `<get>` and `<get-config>` are represented by `GET` requests.
- `<edit-config>` with `create` operation types is represented by `POST` requests.
- `<copy-config>` with `create/replace` operation types is represented by `PUT` requests.
- `<edit-config>`, where the goal of the operation is to change information, is represented by `PATCH` requests.
- `<edit-config>`, where the goal of the operation is to delete information, is represented by `DELETE` requests.

You can find a complete mapping of NETCONF operations to their RESTCONF equivalent in *Section 4* of **Request for Comments** (**RFC**) *8040*. You can find a full link to the RFC here: `https://datatracker.ietf.org/doc/html/rfc8040`.

Making HTTP requests using the requests module in Python

As we have seen, the HTTP protocol is a plain text protocol, and while we could implement this protocol by hand, that is very cumbersome. We will thus use the `requests` module. This module abstracts the process of requesting and handling the response from an HTTP server. In this first recipe, we will see how to do an initial request against a network device that has RESTCONF enabled. This recipe thus also serves as a verification that RESTCONF is properly enabled on your device.

Getting ready

Open your code editor and start by creating a file called `make_request.py`. Next, navigate in your terminal to the same directory in which you just created the `make_request.py` file.

How to do it...

Follow these steps to make your first HTTP request using Python and the `requests` module:

1. Import the `requests` module:

```
import requests
```

2. Create a `requests` session:

```
s = requests.Session()
```

3. Specify the connection information for your device:

```
host = "<insert your host here>"
rest_path = "restconf"
port = <insert your port here>
user = "<insert your username here>"
password = "<insert your password here
```

4. Pass the previously specified authentication information into the session and disable certificate verification if your device uses self-signed certificates when serving the RESTCONF endpoint via **HTTP Secure** (**HTTPS**):

```
s.auth = (user, password)
s.verify = False
```

5. Update the headers to specify that we are sending JSON-formatted YANG data and will accept JSON-formatted YANG data:

```
s.headers.update({
            "Content-Type": "application/yang-data+json",
            "Accept": "application/yang-data+json"
})
```

6. Put together the **Uniform Resource Locator** (**URL**) for our RESTCONF endpoint:

```
url = f"https://{host}:{port}/{rest_path}/"
```

7. Send the request:

```
resp = s.get(url)
```

8. Check the status code and, if the request returned a `200` status code and was thus successful, print out the returned information:

```
if resp.status_code == 200:
            print(resp)
else:
            print(f"Failed to retrieve data. Status code:
{resp.status_code}")
```

9. Finally, run the script by typing the following command in your terminal:

```
python3 make_request.py
```

The output should be a Python dictionary that contains the YANG library version.

How it works...

In order to make an HTTP request using `requests`, we first must import our module. With the module imported, we can then go ahead and create a session. While `requests` also offers us the ability to use functions such as `get()` or `post()` to make the corresponding HTTP requests without creating a session first, the session does have some distinct advantages. With a session, the headers we specify—in this case, the type of content we are sending in the request body and the type of content we would like to receive, as well as our authentication information—are automatically stored between requests. With our user information specified, we can go ahead and first define the authentication details. We assign the session's `auth` variable a tuple that contains our username and password. We also set the `verify` flag to `False`. This tells `requests` to not verify the HTTPS certificates. Since most network devices use self-signed certificates to serve their RESTCONF endpoints over, this is required for the request to not abort with a self-signed certificate error.

We can then update our session's header information by using the `update()` function on our `headers` variable of our previously created `session` object. With the headers specified and stored in our session, we can then go ahead and put the URL for our RESTCONF endpoint together. In this example, we are just sending a request to the main RESTCONF endpoint that exposes some basic information. By sending a request to this endpoint and retrieving a valid response, we also verify that RESTCONF is configured properly and that our login credentials are valid. With our URL specified, we can then go ahead and send the actual request. This is done by using the session's `get()` method that we provide with the URL of the endpoint. This method returns a `Response` object that, among other things, contains a `status_code` member variable that contains the status code of our HTTP response. We can check this status code to ensure that our request was successful.

In this case, we are checking that the status code is 200 and, if so, use the json() function of our Response object to retrieve the JSON-formatted data that was sent in the response body as a parsed Python dictionary. If the request did not return a status code of 200 and was thus not successful, we print out an error message that includes the HTTP status code, which can give us some information about what the problem could be.

There's more...

In the recipe, we specified our headers on a per-session basis. This is great for use cases where you make multiple requests to the same device and do not want to always specify authentication information, for example. We can also pass additional headers on a per-request basis. The following code snippet shows how you can specify these additional headers when making a GET request using the get() function. This works equally for all other request types and their associated functions:

```
import requests
s = requests.Session()
s.headers.update({
        "Content-Type": "application/json"
})
url = f"https://{host}:{port}/{rest_path}/"
addt_headers = {
        "Accept": "application/yang-data+json"
}
resp_one = s.get(url, headers=addt_headers)
resp_two = s.get(url)
```

Here, we are specifying a Content-Type header on the session and creating an additional dictionary with an Accept header that we can pass on to a request using a headers parameter. The first request, which results in resp_one, would thus have both the Content-Type and Accept headers set, while the second request, which results in resp_two, would only have the Content-Type header that was set on the session we used to make this request.

In this recipe, we only showed a `GET` request and its corresponding method, but `requests` also allows us to create and send all the other HTTP methods as well. Here is a list of the important HTTP methods and their corresponding `requests` functions:

- `POST` requests can be performed by using the `post()` function. This function importantly has a `json` keyword argument that allows us to specify a Python dictionary that should be sent as a JSON-formatted payload on the resulting request. A `POST` request that sends a dictionary called `payload` to the server would thus look like this:

```
payload = {
            "some": "data"
}
resp = s.post(url, json=payload)
```

- `DELETE` requests can be performed using the `delete()` function.
- `PATCH` requests can be performed using the `patch()` function. Similar to the `post()` function, they support the `json` keyword argument to specify JSON-formatted data to be sent.
- `PUT` requests can be performed using the `put()` function. Just as with the `post()` and `patch()` functions, this function also supports the `json` keyword argument to specify a payload.

And while we will use sessions throughout this chapter to send requests, you can also send HTTP requests without creating a session first. The `requests` module, just as with the `session` object we created, offers functions for all the different HTTP methods. The next code snippet shows how you can send an HTTP `GET` request without creating a `session` object first:

```
import requests
resp = requests.get("https://packtpub.com")
print(resp.text)
```

This works equally for `POST`/`PATCH`/`PUT`/`DELETE` requests by using the `post()`/`patch()`/`put()`/`delete()` functions.

We have seen that web servers might send redirect instructions (using `3XX` status codes) as a response. By default, requests will follow these redirects for all types of requests except for `HEAD` requests. We can go ahead and change that behavior if we desire by using an `allow_redirects` flag, as shown in the following code snippet:

```
url = "https://packtpub.com"
resp = requests.get(url, allow_redirects=False)
```

`HEAD` requests are used to retrieve the meta-information of a resource without actually retrieving the resource itself. The response to a `HEAD` request should be the same as if you were doing a `GET` request to the specified resource, except that the `HEAD` request will not include a response body like the result of a `GET` request would. `HEAD` requests can thus be used to retrieve information about a response without going through and downloading the entire content of a request.

Retrieving all interfaces of a device using RESTCONF and requests

A sort of `hello world`-type script and a workflow we have seen with different tools throughout this book is the ability to retrieve a list of interfaces currently configured on the device. This can also be achieved using RESTCONF, and in this recipe, we will see how to use the YANG modules that specify our interfaces to retrieve the configuration data present on our device.

Getting ready

Open your code editor and start by creating a file called `retrieve_interfaces.py`. Next, navigate in your terminal to the same directory in which you just created the `retrieve_interfaces.py` file.

How to do it...

Follow these steps to retrieve the interfaces from your RESTCONF-enabled network device:

1. Import the required modules and set up a `requests` session that can interact with the device. We will also import the *prettyprint* standard library to get a nicely formatted output later. For a detailed step-by-step guide on how to connect to your device, please refer to the *Making HTTP requests using the requests module in Python* recipe of this chapter:

```
import requests
import pprint
```

```
s = requests.Session()
host = "<insert your host here>"
rest_path = "restconf"
port = <insert your port here>
user = "<insert your username here>"
password = "<insert your password here>"
s.auth = (user, password)
s.verify = False
s.headers.update({
            "Content-Type": "application/yang-data+json",
            "Accept": "application/yang-data+json"
})
```

2. Put together the URL that we want to send our request to. In this example, we want to query the interfaces that are described by the `ietf-interfaces` YANG module:

```
url = f"https://{host}:{port}/{rest_path}/ietf-
interfaces:interfaces"
```

3. Send the request using our session setup:

```
resp = s.get(url)
```

4. Check the status code of the returned response, and if the response was successful, use *prettyprint* to print out the resulting JSON-formatted data:

```
if resp.status_code == 200:
            pprint.pprint(resp.json())
else:
            print(f"Failed to retrieve data from device.
Status code: {resp.status_code}")
```

5. Finally, run the script by typing the following command in your terminal:

```
python3 retrieve_interfaces.py
```

The resulting output should be a JSON-formatted dictionary that includes entries for all the interfaces configured on your device, including their name, type, and information such as the configured **Internet Protocol** (**IP**) addresses.

How it works...

We first import the `requests` library as well as the *prettyprint* module from the standard library. We then specify the connection variables such as `hostname`, `port`, and the username/password combination used to authenticate ourselves on the device. With the details specified, we can create a `requests` session and define our headers, as well as authentication details, on it. For a more detailed description of how this setup part works, please refer to the *Making HTTP requests using the requests module in Python* recipe of this chapter.

With our session established, we can specify the YANG module we want to query. In this example, we want to retrieve the data for our interfaces, so we are querying the `ietf-interfaces` YANG module and, within that, we are looking for the `interfaces` container. With our URL pieced together, we can go ahead and send the request by using our session. Since we want to retrieve information, we are sending a `GET` request using the `get()` function. Once our request has finished, we check the status codes for error codes and, if the response indicates success by returning a `200` status code, we can go ahead and print out the retrieved information. The information that we get returned is in JSON format. The `Response` object of `requests` offers a handy `json()` function that we can use to automatically parse the JSON-formatted data in the response body to a Python dictionary.

There's more...

In this recipe, we queried our device for information on *all* the interfaces present on our device. However, sometimes we only want to retrieve information about one single interface. This is especially true when talking about requests that apply changes. As is specified by the HTTP protocol and is common practice in REST APIs, this specification should be in the URL. In YANG, we have the concept of leaves. In our previous example, every interface is a leaf of the container interfaces. We can specify one of these leaves by specifying the parameter—for example, the interface name, that we want to check for. Let's say you want to query only the information of the `GigabitEthernet2` interface. You can do that by changing the URL, as follows:

```
url = f"https://{host}:{port}/{rest_path}/ietf-interfaces:inter
faces/?interface=GigabitEthernet2
resp = s.get(url)

if resp.status_code == 200:
    pprint.pprint(resp.json())
```

```
else:
        print(f"Failed to retrieve data from device. Status code:
{resp.status_code}")
```

With the URL specified in the preceding snippet, you might be wondering how you can determine the URL that you have to use based on a YANG module you found for your device. First, you have a base URL. This base URL contains the following:

- The **protocol**—Usually, this will be `https://`.
- The **host**—In our recipes, we are always including them from our `host` variable.
- The **port** that your device is serving RESTCONF on. In our recipes, we are always including this information from the `port` variable.
- The **entry point** for your RESTCONF endpoints. In our recipes, we are specifying this using the `rest_path` variable. This entry point can vary depending on the vendor.

With our `base` URL done, we next need to specify the YANG module that we want to use. In the previous example, this is the `ietf-interfaces` YANG module that specifies a container, as in the `interfaces` example. You can then specify an individual leaf by providing an identification such as the interface name given in the previous example.

It can also be quite handy to be able to list all YANG modules that are available on your device. Similar to how NETCONF lets you retrieve a list of available modules, you can use RESTCONF to do the same. The following code snippet will print a list of the names of all available modules on the command line. You'll have to have a session created for your device:

```
url = f"https://{host}:{port}/{rest_path}/ietf-yang-
library:modules-state"

resp = s.get(url)

if resp.status_code == 200:
        data = resp.json()
        for m in data['ietf-yang-library:modules-state']
['module']:
                print(f"- {m['name']}")
```

Creating a VLAN using RESTCONF and requests

Now that we have seen how to set up our Python scripts to be able to connect to a network device using RESTCONF and how to carry out some basic information retrieval using RESTCONF, we can now go ahead and configure a service. The final three recipes in this chapter, *Creating a VLAN using RESTCONF and requests*, *Updating a VLAN using RESTCONF and requests*, and *Deleting a VLAN using RESTCONF and requests*, show how to use a YANG module to do typical administrative work such as creating, changing, and then deleting a device configuration.

Let's go ahead and first use RESTCONF and `requests` to create a new VLAN configuration.

Getting ready

Open your code editor and start by creating a file called `create_vlan.py`. Next, navigate in your terminal to the same directory in which you just created the `create_vlan.py` file.

How to do it...

Follow these steps to use RESTCONF and `requests` to create a new VLAN:

1. Import the required modules and set up a `requests` session that can interact with the device. We will also import the *prettyprint* standard library to get a nicely formatted output later. For a detailed step-by-step guide on how to connect to your device, please refer to the *Making HTTP requests using the requests module in Python* recipe of this chapter:

```
import requests
import pprint
s = requests.Session()
host = "<insert your host here>"
rest_path = "restconf"
port = <insert your port here>
user = "<insert your username here>"
password = "<insert your password here>"
s.auth = (user, password)
s.verify = False
s.headers.update({
```

```
            "Content-Type": "application/yang-data+json",
            "Accept": "application/yang-data+json"
})
```

2. With our session connected and set up, we can now go ahead and specify the URL that we want to use to configure it:

```
intf_name = "<insert your interface name here>"
url = f"https://{host}:{port}/{rest_path}/ios-xe-native:native/interface/{intf_name}"
```

3. Next, we need to specify the data that we want to send to our RESTCONF server on the device:

```
payload = {
            "name": "3.10",
            "encapsulation": {
                  'dot1Q": {
                        'vlan-id': 10
                  }
            }
}
```

4. With our data and URL specified, we can use a `POST` request from our session to carry out the operation:

```
resp = s.post(url, json=payload)
```

5. Check the status code of the returned response, and if the response was successful, use *prettyprint* to print out the resulting JSON-formatted data:

```
if resp.status_code == 200:
            pprint.pprint(resp.json())
else:
            print(f"Failed to retrieve data from device. Status code: {resp.status_code}")
```

6. Finally, run the script by typing the following command in your terminal:

```
python3 create_vlan.py
```

How it works...

We first import the `requests` library as well as the *prettyprint* module from the standard library. We then specify the connection variables such as `hostname`, `port`, and the username/password combination used to authenticate ourselves on the device. With the details specified, we can create a `requests` session and define our headers, as well as authentication details, on it. For a more detailed description of how this setup part works, please refer to the *Making HTTP requests using the requests module* recipe of this chapter.

With our session established, we can go ahead and specify the information we would like to specify on our interface. With the URL, we are defining on which interface, indicated by the `intf_name` variable, we want to change the `dot1Q` encapsulation. We can then specify all our configuration information in a Python dictionary that contains the keys that are specified as required by the YANG module.

With our data and URL defined, we can then go ahead and use the `POST` request functionality of our session to send the data to our device for configuration.

Updating a VLAN using RESTCONF and requests

Now that we have programmatically created a part of our device configuration using RESTCONF (see the *Creating a VLAN using RESTCONF and requests* recipe), let's request it this time for the next step. Using RESTCONF and the `PATCH` method, we can use a similar data structure to what we have seen when creating the configuration initially, to change the configuration that we previously created.

Let's go ahead and change our previously created VLAN on our interface using RESTCONF and `requests`.

Getting ready

Open your code editor and start by creating a file called `change_vlan.py`. Next, navigate in your terminal to the same directory in which you just created the `change_vlan.py` file.

How to do it...

Follow these steps to define the necessary information to change a previously created VLAN:

1. Import the required modules and set up a `requests` session that can interact with the device. We will also import the *prettyprint* standard library to get a nicely formatted output later. For a detailed step-by-step guide on how to connect to your device, please refer to the *Making HTTP requests using the requests module in Python* recipe in this chapter:

```
import requests
import pprint
s = requests.Session()
host = "<insert your host here>"
rest_path = "restconf"
port = <insert your port here>
user = "<insert your username here>"
password = "<insert your password here>"
s.auth = (user, password)
s.verify = False
s.headers.update({
            "Content-Type": "application/yang-data+json",
            "Accept": "application/yang-data+json"
})
```

2. With our session connected and set up, we can now go ahead and specify the URL that we want to use to configure it:

```
intf_name = "<insert your interface name here>"
url = f"https://{host}:{port}/{rest_path}/ios-xe-
native:native/interface/{intf_name}"
```

3. Next, we need to specify the data that we want to change on our interface. We need to send that data to our RESTCONF server on the device:

```
payload = {
            "name": "3.15",
            "encapsulation": {
                  'dot1Q": {
                        'vlan-id': 15
```

```
                    }
              }
  }
```

4. With our data and URL specified, we can use a `POST` request from our session to carry out the operation:

```
resp = s.patch(url, json=payload)
```

5. Check the status code of the returned response, and if the response was successful, use *prettyprint* to print out the resulting JSON-formatted data:

```
if resp.status_code == 200:
            pprint.pprint(resp.json())
else:
            print(f"Failed to retrieve data from device.
Status code: {resp.status_code}")
```

6. Finally, run the script by typing the following command in your terminal:

```
python3 change_interfaces.py
```

How it works...

We first import the `requests` library as well as the *prettyprint* module from the standard library. We then specify the connection variables, such as `hostname`, `port`, and the username/password combination used to authenticate ourselves on the device. With the details specified, we can create a `requests` session and define our headers, as well as authentication details, on it. For a more detailed description of how this setup part works, please refer to the *Making HTTP requests using the requests module in Python* recipe of this chapter.

With our session established, we can go ahead and define the information we want to change on our interface. With RESTCONF, we can use a Python dictionary to specify the information that needs to be changed on our interface. We first define a target resource by specifying the URL that we will send our configuration payload to. In this example, we are changing the configuration of an interface, and thus we use the native `interface` container. We then specify the interface that we want to change in our `intf_name` variable. With our target specified in the URL, we can go ahead and change the configuration. To do so, we use the same dictionary structure that we have already seen for creating the interface configuration. The variables that need to be specified in this can be found in the YANG module itself.

With our data and URL defined, we can then go ahead and use the `PATCH` request functionality of our session to send the data to our device for configuration. We then check the status code to make sure that our operation was successful and, if that is the case, print out the returned response from our device.

There's more...

You might be wondering when to use the `PATCH` and `PUT` HTTP methods. If you recall the comparison between NETCONF and RESTCONF in the *Revisiting HTTP's request-response model and RESTCONF principles* section, you'll know that both allow you to change a configuration. `PATCH` should be used when you are only changing part of the configuration. `PUT`, on the other hand, is used to replace an entire object with a new object.

Deleting a VLAN using RESTCONF and requests

With the creation (see the *Creating a VLAN using RESTCONF and requests* recipe) and change of a device configuration (see the *Updating a VLAN using RESTCONF and requests* recipe) taken care of, we need to have a look at the final operation we have thus far neglected: deleting resources that are present in our device configuration.

Let's now go ahead and delete our previously created and then changed VLAN programmatically.

Getting ready

Open your code editor and start by creating a file called `delete_vlan.py`. Next, navigate in your terminal to the same directory in which you just created the `delete_vlan.py` file.

How to do it...

Follow these steps to delete a VLAN using RESTCONF and `requests`:

1. Import the required modules and set up a `requests` session that can interact with the device. We will also import the *prettyprint* standard library to get a nicely formatted output later. For a detailed step-by-step guide on how to connect to your device, please refer to the *Making HTTP requests using the requests module in Python* recipe:

```
import requests
import pprint
```

```
s = requests.Session()
host = "<insert your host here>"
rest_path = "restconf"
port = <insert your port here>
user = "<insert your username here>"
password = "<insert your password here>"
s.auth = (user, password)
s.verify = False
s.headers.update({
            "Content-Type": "application/yang-data+json",
            "Accept": "application/yang-data+json"
})
```

2. With our session connected and set up, we can now go ahead and specify the URL that we want to use to delete our VLAN. This URL will define the target device:

```
intf_name = "<insert your interface name here>"
url = f"https://{host}:{port}/{rest_path}/ios-xe-native:native/interface/{intf_name}"
```

3. In addition to the target interface, we also need to specify the target VLAN that we want to delete. In our example in the previous recipe, *Updating a VLAN using RESTCONF and requests*, this VLAN was called `3.10`:

```
target_vlan = "<insert target vlan here>"
url = f"{url}/{target_vlan}"
```

4. With our data and URL specified, we can use a `DELETE` request from our session to invoke a `delete` operation on our device. Notice that the `delete` operation does not take any arguments or additional data:

```
resp = s.delete(url)
```

5. Check the status code of the returned response and, if the response was successful, use *prettyprint* to print out the resulting JSON-formatted data:

```
if resp.status_code == 200:
            pprint.pprint(resp.json())
else:
            print(f"Failed to retrieve data from device. Status code: {resp.status_code}")
```

6. Finally, run the script by typing the following command in your terminal:

```
python3 delete_interfaces.py
```

How it works...

We first import the `requests` library as well as the *prettyprint* module from the standard library. We then specify connection variables such as `hostname`, `port`, and the username/password combination used to authenticate ourselves on the device. With the details specified, we can create a `requests` session and define our headers, as well as authentication details, on it. For a more detailed description of how this setup part works, please refer to the *Making HTTP requests using the requests module in Python* recipe.

With our session established, we can go ahead and define the resource—in our case, the VLAN on our specified interface—that we want to delete. Resources in RESTCONF are always specified by their URL. In our Python script, we first specify the YANG module (see the *There's more...* section in the *Retrieving all interfaces of a device using RESTCONF and requests* recipe for an overview of how to identify the correct URL) and then the interface (specified in the `intf_name` variable). We then amend this information with the VLAN that we want to delete. With our resource uniquely identified in the URL, we can use the `delete()` function to issue a `DELETE` request to our RESTCONF server on the device. We then check if the request was carried out correctly and print out the status code if that's not the case.

9 Consuming Controllers and High-Level Networking APIs with requests

Why do we have to always connect to each of our devices individually to carry out some operation? This is a question you might have asked yourself throughout this book. And while device-by-device configuration remains a common way to interact with networking device configuration, controller-based approaches, where a central component coordinates and controls all devices on a network, are on the rise. And these control components usually offer us an **application programming interface** (**API**) that we can use to interact with the controller. The controller then goes out to all the devices we have registered on it and retrieves the required information or applies the required configuration change, without us having to explicitly apply this configuration change to all our devices. The controller takes over the job of connecting to each of the devices.

The common architectural style these APIs use is **REST**, short for **REpresentational State Transfer**. REST builds on top of the **Hyper Text Transfer Protocol** (**HTTP**) and has, since its inception in the early 2000s, become the de facto standard for building web APIs. In this chapter, we will have a look at one specific REST API, known as the Dashboard API, which can be used to interact with the Cisco Meraki family of devices. To interact with this API, we are going to use the `requests` package.

This chapter will show the principles of interacting with a REST API using the Cisco Meraki Dashboard API as an example. These principles can be applied to any other REST API. Additionally, many REST APIs provide a Python package that wraps the API calls. This is also the case for Meraki, and you'll see how to do the same requests you did manually with the Meraki SDK, which is the recommended way of doing things for production usage.

In this chapter, we will be covering the following recipes:

- Authenticating web requests
- Storing authentication metadata between requests sessions
- Retrieving a list of Meraki networks
- Retrieving usage details and connected clients for a Meraki network
- Rebooting a Meraki device
- Retrieving channel usage for your Meraki access point
- Updating the switch port configuration of a Meraki device
- Deleting the QoS rules on a Meraki device
- Using webhooks to programmatically react to an **Access Point** (**AP**) going down

Technical requirements

For this chapter and the remainder of this book, you'll need to install Python. Specifically, you'll need a Python interpreter that's version 3.6.1 or higher. This book makes use of the language constructs of Python 3, which means it is incompatible with Python 2.x. Additionally you'll need to install the `requests` package. You can install the newest version of `requests` using `python3 -m pip install requests`. At the time of writing, the current version is version 2.25.1.

If you want to follow the *There's more...* sections, you'll also need the Meraki SDK. To install the newest version of the package, you can use `python3 -m pip install meraki`. At the time of writing, the current version is version 1.7.2.

You will also need a code editor. Popular choices include Microsoft Visual Studio Code or Notepad++. Additionally, you'll need a Meraki organization that has API access enabled. You can find a guide on how to do this for your organization at `https://documentation.meraki.com/General_Administration/Other_Topics/Cisco_Meraki_Dashboard_API`.

You can view this chapter's code in action here: `https://bit.ly/3fWD2rc`

Authenticating web requests

In this chapter, we will learn how APIs handle authentication when they're queried. While there are some APIs out there that do not require any sort of authentication, most APIs will require some sort of user-provided authentication in order to associate a request with a user and verify that that user is actually who they claim to be. This also helps ensure they have the right set of permissions to access the resource that is being requested.

The most common way to carry out this authentication is by using an `Authorization` header. You may want to refer to *Chapter 8, Configuring Devices Using RESTCONF and requests*, the *Revisiting the HTTPS request-response module* recipe, for a more detailed recap of how HTTP requests work, but in summary, headers are additional meta information that can be sent by a client to the server. The header is a key-value pair, where in our example the key is **Authorization** and the value is the secret that authenticates us against the API server. This secret, commonly referred to as an **access token**, or simply **token**, is used to authenticate and authorize requests on behalf of a user. It is equal to a username/password combination. There are different types of tokens, but they generally fall into one of two categories:

- **Tokens with an indefinite validity**: Once generated, these tokens stay valid until they are revoked by the user. Meraki uses these types of indefinite valid tokens, and they can usually be generated in the web interface of the web services of the API you want to interact with.
- **Tokens with a limited validity**: These are tokens that only have a limited time span in which they are valid. This timespan can range from a few minutes to several days or weeks. However, once the token is past its validity date, it can no longer be used. While tokens with indefinite validity can usually be obtained using a web interface, APIs that use limited tokens usually offer an API endpoint that we can send a username/password combination to, in order to obtain the token. An example of an API that uses such an approach would be Cisco DNA Center.

While the `Authorization` header is the most commonly used field name to specify authorization information, there is no standard on this, and an API service can choose to define a different header field to use. Notably, Cisco Meraki uses the `X-Cisco-Meraki-API-Key` header field instead of the `Authorization` field.

Passing a username-password combination as authentication data in a request

First, let's have a look at how to pass a username-password combination to an API service using requests. Username-password authentication, sometimes also referred to as *Basic Auth*, uses the `Authorization` header in a request and specifies the username and password as the value of the header field. These credentials are not transferred in plain text; instead, the username and password, separated by a colon sign, are base64-encoded and prepended with the **Basic** keyword. This keyword indicates to the server that the HTTP Basic auth schema is being used. Therefore, the final `Authorization` header for a request that uses the `root` username and the `test` password as credentials would look like this:

```
Authorization: Basic cm9vdDp0ZXN0
```

Here, `cm9vdDp0ZXN0` is the base64-encoded version of `root:test`.

While we could use Python's built-in modules to create a base64-encoded string based on a username-password combination and then manually set the header to the correct values, `requests` offers a convenient way of passing a username/password combination to a request and automatically getting the correctly encoded headers set. The following code imports `requests` as well as the `HTTPBasicAuth` model from the requests auth subpackage, and then uses them to authenticate a `GET` request to `packt.com` using the `root` username and `test` password. Once the request is complete, it prints out the status code of the request we just completed:

```
import requests
from requests.auth import HTTPBasicAuth

auth = HTTPBasicAuth("root", "test")

response = requests.get("https://packt.com", auth=auth)
print(response.status_code)
```

While this is a common way of authenticating requests, some APIs may also require you to send the username and password in the body of a `POST` request. We will discuss how to `POST` data to an API later in this chapter. Check the authentication details of your API docs to see how the access tokens are obtained. Some APIs might even allow you to authenticate all your requests using `HTTP Basic Auth`, in which case you can use the preceding method to just use the API itself without having to obtain an access token.

`requests` provides you with a handy shorthand when using HTTP Basic Auth. While you can import the `HTTPBasicAuth` class and create a new `Auth` object, you can also just pass a tuple containing the username/password combination. Since you may encounter both versions in sample scripts, I have opted to include both here. With this new shorthand, the preceding snippet can also be written like so:

```
import requests
auth = ("root", "test")
response = requests.get("https://packt.com", auth=auth)
print(response.status_code)
```

Passing a token to a request using custom header fields

But what about APIs that, instead of letting us specify a username and password to authenticate, require an authentication token? In requests, we can specify custom header fields for every request we make. Each of the HTTP request methods, such as `GET`, `POST`, and `DELETE`, have a corresponding method in the `requests` package, and each of these methods accepts the `headers` keyword argument. This argument takes a dictionary and passes every key-value pair of that dictionary as an additional header field.

The following code imports the `requests` module and then specifies additional headers in a dictionary called `headers`. Once the headers have been specified, they are passed on to a `GET` request to `packt.com` and the response code is printed:

```
import requests
headers = {
      'Authorization': f"Bearer thiswouldbeatoken"
}

response = requests.get("https://packt.com", headers=headers)
print(response.status_code)
```

As you can see, we specify our `Authorization` header field in our `headers` dictionary and then pass this information on to the request method, which in this case is `get()`.

So far, we have always set our credentials, be it the username/password combination or our dummy token from the preceding example, explicitly in the source code. However, this is bad practice in production. Your authentication tokens are as valuable as a password, so writing them into the source code is highly problematic. You may accidentally send the source code file, including your credentials, to a colleague or check it into a version control system. So, where should we put our tokens? Instead of putting them into the source code directly, we can use the concept of environment variables.

On Linux/macOS, you can use the following command to specify an environment variable called `ACCESS_TOKEN` with a value of `THIS_IS_A_TOKEN`:

```
export ACCESS_TOKEN=THIS_IS_A_DUMMY_TOKEN
```

On Windows, you would use the `set` command:

```
set ACCESS_TOKEN=THIS_IS_A_DUMMY_TOKEN
```

This will set the environment variable for our current terminal session. Then, within our Python script, we can use the built-in `os` package to retrieve the variable. The following code shows how to read the `ACCESS_TOKEN` environment variable and print out its value:

```
import os
access_token = os.environ.get("ACCESS_TOKEN")

print(f"The access token is {access_token}"
```

Going forward, we are going to use this to specify the access token that will be used to authenticate against the Meraki Dashboard API.

Storing authentication metadata between requests using sessions

In the previous chapter, you learned how to specify header fields for a request. But if we are sending multiple requests to the same API, and thus require each of these requests to have the same authentication information, we will have to pass the `headers` dictionary on every single request. While we could do this, it requires additional effort and is very prone to errors since we might forget to pass the headers in one of our requests. To make our life easier and our code more maintainable, the `requests` module provides us with a way of storing header fields between requests by creating a `Session` object, and then setting the headers on it. Let's see how we can use such a `Session` object to specify the headers required for authenticating against the Meraki Dashboard API, and then use a `GET` request to retrieve all the organizations associated with the user this token belongs to.

Getting ready

Open your code editor and start by creating a file called `authenticate.py`. Next, navigate your terminal to the same directory that you just created the `authenticate.py` file in.

In your terminal, you'll also have to set an environment variable called `MERAKI_DASHBOARD_API_KEY`, whose value is the API key for your Meraki organization. You can refer to the *Authenticating web requests* section for an introduction on how to set environment variables on different operating systems.

How to do it...

Follow these steps to authenticate your request against the Meraki API using a `Session` object and retrieve all organizations:

1. Import the built-in `os` and `sys` modules, as well as our `requests` module:

```
import os
import sys
import requests
```

2. Retrieve the API key from our environment variable and check that it is not `None`. We must also specify the base URL of our API:

```
base_url = "https://api.meraki.com/api/v1"
key = os.environ.get("MERAKI_DASHBOARD_API_KEY", None)
if key is None:
            print("Please provide an API key. Aborting.")
            sys.exit(-1)
```

3. Create a new `requests` `Session` object:

```
sess = requests.Session()
```

4. On that `Session`, update the headers so that they include our authorization details:

```
sess.headers.update({
            " X-Cisco-Meraki-API-Key": key
})
```

5. Create the URL where we will receive a list of all our organizations and send a `GET` request using our session:

```
url = f"{base_url}/organizations"
resp = sess.get(url)
```

6. Check the status code of our response and if it's `200`, indicating that the request was successful, parse the JSON the API has sent us into a dictionary and print out all the organizations:

```
if resp.status_code == 200:
            data = resp.json()
            for org in data:
                  print(org['name'])
else:
            print(f"Bad response. Status code: {resp.
status_code}")
```

7. Execute the script by running the following command in your terminal:

```
python3 auth.py
```

How it works...

First, we imported three modules: the built-in `sys` and `os` modules from the standard library, as well as our `requests` module. We then specified the base URL of our API, which in this case was `https://api.meraki.com/api/v1`, and retrieved the access token from the environment. To do this, we used a dictionary called `environ` that is specified by the `os` module. On that dictionary, we used the `get()` function to either retrieve the value stored in the environment under the `MERAKI_DASHBOARD_API_KEY` variable or `None`. We then checked that we had retrieved something from the environment variable. If that didn't happen, we would have to stop executing our script by using the `exit()` function provided by the `sys` module. With the required authentication details retrieved, we created a new `Session` object and updated the session's headers to include our `Authorization` header field and the previously retrieved API key as the value.

Next, we specified the URL of the resource, which in this case was the organization's, that we wanted to retrieve and sent a `GET` request. Instead of using the `get()` function provided by the `requests` module itself (`requests.get()`), we called the `get()` function of our session to retrieve the data from the previously specified URL using the previously specified headers, and then stored the response in our `resp` variable. We then checked the status code of our response and, if it was a `200`, indicating a successful request, we retrieved the JSON-encoded data provided in the body of our response. We then had `requests` parse this into a Python dictionary by using the response's `json()` function. This returns a list of dictionaries that represent our different organizations. The keys of each of these dictionaries are the names of the organizations that we can print out.

There's more...

We can also use the Meraki SDK to achieve the same outcome, without having to manually do the requests ourselves. You'll have to have the same environment variable (`MERAKI_DASHBOARD_API_KEY`) set:

```
import meraki

dashboard = meraki.DashboardAPI()

orgs = dashboard.organizations.getOrganizations()

for org in orgs:
    print(org['name'])
```

As you can see, the Meraki SDK takes care of putting together the correct URL based on our function call. It also takes care of sending and retrieving the request for us.

When checking for status codes, instead of comparing the value from the response to the number itself, a response object obtained from `requests` can also check that by comparing the status code and checking that it is below `400`. The response includes a Boolean property of `ok`, which will be true if the status code is less than `400`:

```
if resp.ok:
    data = resp.json()
```

Retrieving a list of Meraki networks

Meraki organizes all its information in a hierarchy, where a device belongs to a network and a network, in turn, belongs to an organization that can have multiple networks. A common task when dealing with REST APIs is to retrieve all the information that is currently stored on the system. In this example, we will see how to retrieve all the networks that are registered to all our organizations. This will require us to make multiple HTTP requests.

Getting ready

Open your code editor and start by creating a file called `networks.py`. Next, navigate your terminal to the same directory that you just created the `networks.py` file in.

In your terminal, you'll also have to set an environment variable called `MERAKI_DASHBOARD_API_KEY`, whose value is the API key for your Meraki organization. You can refer to the *Authenticating web requests* recipe for an introduction on how to set environment variables on different operating systems.

How to do it...

Follow these steps to authenticate your request against the Meraki API and retrieve a list of networks for each of your organizations:

1. Set up a `requests` session using the API key stored in your environment. Please refer to the *How to do it...* section of the *Storing authentication and metadata in requests sessions* recipe for a detailed explanation of each of these steps:

```
import os
import sys
import requests

base_url = "https://api.meraki.com/api/v1"
key = os.environ.get("MERAKI_DASHBOARD_API_KEY", None)
if key is None:
            print("Please provide an API key. Aborting.")
            sys.exit(-1)
sess = requests.Session()
sess.headers.update({
            " X-Cisco-Meraki-API-Key": key
})
```

2. With our session established, we must retrieve all the organizations associated with our account:

```
orgs_url = f"{base_url}/organizations"
resp_orgs = sess.get(orgs_url)
```

3. If our request was successful, and thus returned with a status code of `200`, we can iterate over all the organizations that were returned:

```
if resp_orgs.status_code == 200:
    for org in resp_orgs.json():
```

4. Then, we can extract the organization ID and create a new URL to request all networks for the organization specified by the ID, as well as print out the name of the organization:

```
        org_id = org['id']
        print(f"Retrieving networks for {org['name']}")
        url = f"{base_url}/organizations/{org_id}/
networks"
```

5. Next, we will send a request to gather all the networks and, if that request is successful, iterate over all the networks and print out the networks' names:

```
        resp = sess.get(url)

        if resp.status_code == 200:
            for ntwrk in resp.json():
                print(f"- {ntwrk['name']}")
```

6. Finally, we must make sure that we handle any non-successful status codes both from our network and organization requests:

```
        else:
            print(f"Failed to retrieve network. Status
code: {resp.status_code}")
else:
    print(f"Failed to retrieve orgs. Response code:
{resp_orgs.status_code}")
```

7. Execute the script by running the following command in your terminal:

```
python3 networks.py
```

You should see a list of all your organizations, as well as the names of all your networks.

How it works...

We started by creating an authenticated requests session by retrieving the Meraki API key from our environment and updating the header fields that are sent with each request. Please refer to the *How it works...* section of the *Storing authentication and metadata in requests sessions* recipe for a more detailed explanation of how this works.

With our session established, we retrieved all the organizations. We needed to do this since the Meraki hierarchy puts networks under an organization, so, using our authenticated session, we performed a `GET` request against the `/organizations` endpoint. If this request is successful, it will have a status code of `200` and we can iterate over all the organizations associated with our account. Mainly, we wanted to print out the name of our organization and extract the ID. With this ID, we pieced together the URL for requesting the networks associated with each ID. They follow the format of `/organizations/<org_id>/networks`. We then used our session to perform a request for all the networks and, if the request was successful, parsed the returned JSON list into a list of Python dictionaries and, for each dictionary representing a network in that organization, printed out the name of the network. Finally, we needed to put in our debug information in case one of our requests fails. In this example, we printed out a message specifying which request – either a request for an organization or for a network that has failed – and what the response code was.

There's more...

We can also use the Meraki SDK to achieve the same outcome, without having to manually do the requests ourselves. You must have the same environment variable (`MERAKI_DASHBOARD_API_KEY`) set:

```
import meraki

dashboard = meraki.DashboardAPI()
orgs = dashboard.organizations.getOrganizations()

for org in orgs:
        networks = dashboard.organizations.
getOrganizationNetworks(org['id'], total_pages='all')
        for ntwrk in networks:
                print(f"- {ntwrk['name']}")
```

As you can see, the Meraki SDK takes care of putting together the correct URL based on our function calls. Here, we pass the required parameter, which in this case is the ID of the organization of our `getOrganizationNetworks()` function.

Retrieving usage details and connected clients for a Meraki network

How many clients do you currently have on your network? What are their MAC addresses? How much data are they sending or receiving? When did the client first connect and when was the client last online? These kinds of questions lend themselves to automated report creation. In this recipe, we will explore how to extract these metrics using our API, and then save them into a CSV file that can be opened with a table calculation program such as Excel for further analysis.

Getting ready

Open your code editor and start by creating a file called `usage.py`. Next, navigate your terminal to the same directory that you just created the `usage.py` file in.

In your terminal, you'll also have to set an environment variable called `MERAKI_DASHBOARD_API_KEY`, whose value is the API key for your Meraki organization. Please refer to the *Authenticating web requests* recipe for an introduction on how to set environment variables on different operating systems.

How to do it...

Follow these steps to authenticate your request against the Meraki API and create a usage report:

1. Set up a `requests` session using the API key stored in your environment. Please refer to the *How to do it…* section of the *Storing authentication and metadata in requests sessions* recipe for a detailed explanation of each of these steps. Additionally, we are importing Python's built-in `csv` module to create our file export with:

```
import os
import sys
import csv

import requests

base_url = "https://api.meraki.com/api/v1"
key = os.environ.get("MERAKI_DASHBOARD_API_KEY", None)
```

```
if key is None:
            print("Please provide an API key. Aborting.")
            sys.exit(-1)
sess = requests.Session()
sess.headers.update({
            " X-Cisco-Meraki-API-Key": key
})
```

2. With our session set up, we can get started with our calls. We will start by setting the ID of the network that we want to query the clients for and put our URL together:

```
network_id = "<insert network id here>"
url = f"{base_url}/networks/{network_id}/clients
```

3. Request all client information and, if successful, retrieve a list of dictionaries containing all the clients from the past 31 days:

```
resp = sess.get(url)
if resp.status_code == 200:
    clients = resp.json()
```

4. Open a new file called `clients.csv` for writing and create a `csv.writer` that we will use to export any information that's retrieved from the API:

```
    with open('clients.csv', 'w') as csvfile:
        out = csv.writer(csvfile, delimiter=',')
```

5. Next, we must export the keys of our clients as the header for our `csv` file:

```
        out.writerow(clients[0].keys())
```

6. Then, we will iterate over each of our client dictionaries and write the specific information into the `csv` file:

```
        for client in clients:
            out.writerow(client.values())
```

7. Lastly, we must react if our status code is not `200`:

```
else:
    print(f"Request failed. Status code: {resp.status_
code}")
```

8. Execute the script by running the following command in your terminal:

```
python3 usage.py
```

Once the script has finished running, you should see a file called `clients.csv` that contains your extracted information.

How it works...

We started by creating an authenticated requests session by retrieving the Meraki API key from our environment and updating the header fields that are sent with each request. Please refer to the *How it works...* section of the *Storing authentication and metadata in requests sessions* recipe for a more detailed explanation of how this works. Additionally, we included the `csv` module from the standard library for exporting purposes.

With the session established, we specified the ID of the network that we wanted to run our report on. You could, of course, used the code in the *Retrieving a list of Meraki networks* recipe to retrieve the IDs of all our networks and run this code snippet for all of them, but in this example, we specified a network ID. With this ID, we can put together the URL. In this case, the URL follows the `/networks/<network_id>/clients` schema. We requested the information on all clients with this URL and, if the request was successful, we opened a new file for writing called `clients.csv`. Then, we passed the newly created file handler into a `csv.writer` object, which we can use to easily write a list of values into the `csv` file. Since the API returned a list of dictionaries, we used the keys of these dictionaries as the header of our `csv` file. Once the header was written to the file, we iterated over each of the clients we retrieved from the API and wrote the required information to the `csv` file.

There's more...

We can also use the Meraki SDK to achieve the same outcome, without having to manually do the requests ourselves. You must have the same environment variable (`MERAKI_DASHBOARD_API_KEY`) set:

```
import csv

import meraki

network_id = "<insert network id here>"
dashboard = meraki.DashboardAPI()
clients = dashboard.networks.getNetworkClients(network_id,
total_pages='all')
```

```
with open('clients.csv', 'w') as csvfile:
        out = csv.writer(csvfile, delimiter=',')
        out.writerow(clients[0].keys())
        for client in clients:
                out.writerow(client.values())
```

As you can see, the Meraki SDK takes care of putting together the correct URL based on our function call. We can then use the same logic that we used for our manual requests to save the data to a `csv` file.

Rebooting a Meraki device

So far, in our requests, we have only retrieved information by using `GET` requests. But what about *changing* something? Creating or changing a resource in a REST API is usually done by sending `POST` or `PATCH/PUT` requests. `POST` requests are typically used when creating a resource or kicking off an action, while `PATCH/PUT` are used when updating a resource. In this recipe, we are going to reboot a Meraki device, based on its serial number, using a `POST` request.

Getting ready

Open your code editor and start by creating a file called `reboot.py`. Next, navigate your terminal to the same directory that you just created the `reboot.py` file in.

In your terminal, you'll also have to set an environment variable called `MERAKI_DASHBOARD_API_KEY`, whose value is the API key for your Meraki organization. Please refer to the *Authenticating web requests* recipe, for an introduction on how to set environment variables on different operating systems.

How to do it...

Follow these steps to authenticate your request against the Meraki API and reboot a device based on its serial number:

1. Set up a `requests` session using the API key stored in the environment. Please refer to the *How to do it...* section of the *Storing authentication and metadata in requests sessions* recipe for a detailed explanation of each of these steps:

```
import os
import sys
import requests

```

```
base_url = "https://api.meraki.com/api/v1"
key = os.environ.get("MERAKI_DASHBOARD_API_KEY", None)
if key is None:
            print("Please provide an API key. Aborting.")
            sys.exit(-1)
sess = requests.Session()
sess.headers.update({
            " X-Cisco-Meraki-API-Key": key
})
```

2. With our session set up, we can specify the serial number of the device we want to reboot and put together the URL based on that serial number:

```
serial = "<Insert device serial here>"
url = f"{base_url}/devices/{serial}/reboot
```

3. Next, we will use the `post()` method of our session to kick off the request and save the response in `resp`:

```
resp = sess.post(url)
```

4. Finally, we will check the status code and print out a different message to the user, based on whether the request was a success or not:

```
if resp.status_code == 202:
            print("Rebooted device successfully.")
else:
            print(f"Failed to reboot device. Status code:
{resp.status_code}")
```

5. Execute the script by running the following command in your terminal:

```
python3 reboot.py
```

You should see a message stating whether the reboot was successful or not. After, your device should start rebooting.

How it works...

We started by creating an authenticated requests session by retrieving the Meraki API key from our environment and updating the header fields that are sent with each request. Please refer to the *How it works...* section of the *Storing authentication and metadata in requests sessions* recipe for a more detailed explanation of how this works.

With our session established, we pieced together the URL. Since we were dealing with devices as our main resource, and these devices are identified by their serial number, the schema we used was `/devices/<serial>/reboot`. So, instead of specifying additional resources, such as clients, for our previous `GET` request to retrieve all the clients of a network, we specified `reboot` as the additional resource. Then, we sent a `POST` request using our session and checked the response for a status code, indicating either success or failure. Note how the API returned a status code of `202` upon success. This status code indicates that our request was successfully received, and that the API is now in the process of restarting the device.

There's more...

We can also use the Meraki SDK to achieve the same outcome, without having to manually do the requests ourselves. You must have the same environment variable (`MERAKI_DASHBOARD_API_KEY`) set:

```
import meraki
serial = "<Insert device serial here>"
dashboard = meraki.DashboardAPI()
dashboard.devices.rebootDevice(serial)
```

As you can see, the Meraki SDK takes care of putting together the correct URL based on our function call. It also takes care of sending and retrieving the request for us, as well as using a `POST` request instead of the `GET` requests we used previously. Notice how the function call changed from starting with `get` to starting with `reboot`, indicating that we are now carrying out an action instead of retrieving information.

Here, we are checking the status code of our response manually against the desired status code of `202`. If we are not interested in the actual status code, we could also use the built-in `ok` property of a response. This property is set to `True` if the status code of a response is below `400`. Therefore, we could write the following:

```
if resp.ok:
      print("Request was succesful")
```

Retrieving channel usage for your Meraki access point

Are your Wi-Fi points using their channels properly? And what is the channel utilization over each radio? We can use a script to retrieve the total utilization for each Wi-Fi point in a Meraki network and print this information out for a user to look at. Meraki allows us to specify the time frame, so we are going to use query parameters to specify a suitable one.

Getting ready

Open your code editor and start by creating a file called `utilization.py`. Next, navigate your terminal to the same directory that you just created the `utilization.py` file in.

In your terminal, you'll also have to set an environment variable called `MERAKI_DASHBOARD_API_KEY`, whose value is the API key for your Meraki organization. Please refer to the *Authenticating web requests* recipe for an introduction on how to set environment variables on different operating systems.

How to do it...

Follow these steps to authenticate your request against the Meraki API and retrieve the total utilization for all your Wi-Fi points in the last 7 days:

1. Set up a `requests` session using the API key stored in your environment. Please refer to the *How to do it...* section of the *Storing authentication and metadata in requests sessions* recipe for a detailed explanation of each of these steps:

```
import os
import sys
import requests

base_url = "https://api.meraki.com/api/v1"
key = os.environ.get("MERAKI_DASHBOARD_API_KEY", None)
if key is None:
            print("Please provide an API key. Aborting.")
            sys.exit(-1)
sess = requests.Session()
sess.headers.update({
            " X-Cisco-Meraki-API-Key": key
})
```

2. With our session established, we can set up the URL. We need a network ID that we want to receive all the information for:

```
network_id = "<Insert network id here>"
url = f"{base_url}/networks/{network_id}/networkHealth/
channelUtilization
```

3. Next, we want to specify the query parameters that will be passed on. We can specify them in a dictionary:

```
params = {
            'timespan': 7
}
```

4. Using our session, we will execute a `GET` request and pass the `query` parameter:

```
resp = sess.get(url, params=params)
```

5. If our request was successful, we will retrieve the list of our APs and then, for each network in an AP, print out the total utilization:

```
if resp.status_code == 200:
            aps = resp.json()
            for ap in aps:
                  print(f"Wifies on AP {ap['serial']}")
                  for e in ap['wifi0']:
                        print(f"{e['utilizationTotal']}")
```

6. Lastly, we need to react to a non-successful response and print out the status code:

```
else:
            print(f"Failed to retrieve utilization. Code:
{resp.status_code}")
```

7. Execute the script by running the following command in your terminal:

```
python3 utilization.py
```

You should see the utilization for each of your AP's Wi-Fi points printed.

How it works...

We started by creating an authenticated `requests` session by retrieving the Meraki API key from our environment and updating the header fields that are sent with each request. Please refer to the *How it works...* section of the *Storing authentication and metadata in requests sessions* recipe for a more detailed explanation of how this works.

With our session established, we could specify our URL. In addition to the path parameter, which in this case was our network ID, we specified some query parameters. These parameters are passed on in the query part of the URL. While we could have added our query parameters to the URL manually, we can also have `requests` do this based on a dictionary by specifying the dictionary and then passing it on to the request function using the `params` argument. In this example, we set the timeframe that we wanted to look back to from the default of 1 day to 7 days. In REST APIs, query parameters are usually used to specify optional parameters such as filters.

With our URL put together, we sent the request, parsed the JSON response, and iterated over all the entries to print the information back to the user. Finally, we checked for non-successful status codes and printed an error message back that included the status code of our failed request.

There's more...

We can also use the Meraki SDK to achieve the same outcome, without having to manually do the requests ourselves. You must have the same environment variable (`MERAKI_DASHBOARD_API_KEY`) set:

```
import meraki

dashboard = meraki.DashboardAPI()
network_id = "<Insert network id here>"
data = dashboard.networks.
getNetworkNetworkHealthChannelUtilization(network_id,
timespan=7, total_pages='all')

for ap in aps:
        print(f"Wifies on AP {ap['serial']}")
        for e in ap['wifi0']:
                print(f"{e['utilizationTotal']}")
```

As you can see, the Meraki SDK takes care of putting together the correct URL based on our function call. You can see that we are passing on additional arguments, such as the timespan in our example, as keyword arguments.

Updating the switchport configuration of a Meraki device

So far, we have retrieved information and sent requests to carry out operations. But what we have not done yet is update a resource. In this recipe, we are going to change the switchport maximum number of sticky MAC addresses allowed on our switch port to `10`.

Getting ready

Open your code editor and start by creating a file called `update.py`. Next, navigate your terminal to the same directory that you just created the `update.py` file in.

In your terminal, you'll also have to set an environment variable called `MERAKI_DASHBOARD_API_KEY`, whose value is the API key for your Meraki organization. Please refer to the *Authenticating web requests* recipe for an introduction on how to set environment variables on different operating systems.

How to do it...

Follow these steps to authenticate your request against the Meraki API and update the switch port configuration:

1. Set up a `requests` session using the API key stored in the environment. Please refer to the *How to do it...* section of the *Storing authentication and metadata in requests sessions* recipe for a detailed explanation of each of these steps:

```
import os
import sys
import requests

base_url = "https://api.meraki.com/api/v1"
key = os.environ.get("MERAKI_DASHBOARD_API_KEY", None)
if key is None:
            print("Please provide an API key. Aborting.")
            sys.exit(-1)
sess = requests.Session()
sess.headers.update({
            " X-Cisco-Meraki-API-Key": key
})
```

2. With our session established, we need a device serial number and the ID of the port to update, such as port `1`. Make sure that this port has been configured with the access policy type; that is, `Sticky MAC allow list`:

```
serial = "<INSERT Serial here>"
port = <Insert port number here>
url = f"{base_url}/devices/{serial}/switch/ports/{port}
```

3. Next, we are going to retrieve the current configuration using a `GET` request:

```
resp = sess.get(url)
```

4. If our `GET` request is successful, we can retrieve the current configuration of our switch port as a Python dictionary and change the value – in our case, the number of allowed sticky mac addresses – and send a `PUT` request to the same URL to change the configuration:

```
if resp.status_code == 200:
            conf = resp.json()
            conf['stickyMacAllowListLimit'] = 10
            resp_update = sess.put(url, json=conf)
            if resp_update.status_code == 200:
                  print("Config updated!")
```

5. Lastly, we need to respond to any of our requests failing:

```
        else:
                  print(f"Failed to update. Status:
{resp_update.status_code}")
else:
            print(f"Failed to get config. Status: {resp.
status_code}")
```

6. Execute the script by running the following command in your terminal:

```
python3 update.py
```

You should see a message indicating that the port configuration has been updated. You should also see the changed limit in your dashboard.

How it works...

We started by creating an authenticated `requests` session by retrieving the Meraki API key from our environment and updating the header fields that are sent with each request. Please refer to the *How it works...* section of the *Storing authentication and metadata in requests sessions* recipe for a more detailed explanation of how this works.

With our session established, we retrieved our switch port configuration. The URL we needed identified the device by its serial number and then specified that we are looking at the port with the specified number. The URL schema here was `/devices/{serial}/switch/ports/{port}`. With the URL put together, we made a request to retrieve the current configuration state, which we could retrieve as a JSON dictionary from the API. With the current configuration retrieved, we applied the change we wanted to make – in this case, the number of sticky MACs allowed – and sent the updated configuration back to Meraki to be applied to our device. Finally, we reacted to any of our requests failing by printing out the status code for further investigation.

Deleting the QoS rules on a Meraki device

The last kind of operation we sometimes need to carry out is deleting a configuration. For REST APIs, this functionality can be achieved by sending a `DELETE` request. In this recipe, we are going to learn how to use a combination of `GET` and `DELETE` requests to delete all QoS rules that were previously present in our specific network.

Getting ready

Open your code editor and start by creating a file called `qos.py`. Next, navigate your terminal to the same directory that you just created the `qos.py` file in.

In your terminal, you'll also have to set an environment variable called `MERAKI_DASHBOARD_API_KEY`, whose value is the API key for your Meraki organization. Please refer to the *Authenticating web requests* recipe for an introduction on how to set environment variables on different operating systems.

How to do it...

Follow these steps to authenticate your request against the Meraki API so that you can retrieve a list of QoS rules for a specific network and then delete them:

1. Set up a `requests` session using the API key stored in your environment. Please refer to the *How to do it...* section of the *Storing authentication and metadata in requests sessions* recipe for a detailed explanation of each of these steps:

```
import os
import sys
import requests

base_url = "https://api.meraki.com/api/v1"
key = os.environ.get("MERAKI_DASHBOARD_API_KEY", None)
if key is None:
            print("Please provide an API key. Aborting.")
            sys.exit(-1)
sess = requests.Session()
sess.headers.update({
            " X-Cisco-Meraki-API-Key": key
})
```

2. With our session set up, we can specify the network ID where we want to delete all our QoS rules and, based on that, specify the URL to retrieve all rules:

```
network_id = "<Insert network id here>"
url = f"{base_url}/networks/{network_id}/switch/qosRules"
```

3. Retrieve all the rules that are present:

```
resp = sess.get(url)
if resp.status_code == 200:
            rules = resp.json()
```

4. Iterate over all the rules and send a `DELETE` request using our session, the network ID, and the ID of the rule we want to delete:

```
       for rule in rules:
            url_del = "{base_url}/networks/{network_id}/
switch/qosRules/{rule['id']}"
            resp_del = sess.delete(url_del)
            if resp_del.status_code == 204:
                 print(f"Deleted QoS rule {rule['id']}")
```

5. Finally, we need to check for failing requests and handle them by displaying the status code to the user for further investigation:

```
            else:
                    print(f"Failed on delete request.
Status: {resp_del.status_code}")
else:
        print(f"Failed to retrieve rules. Status:
{resp.status_code}")
```

6. Execute the script by running the following command in your terminal:

```
python3 qos.py
```

You should see a list of all the QoS rules in your system being deleted.

How it works...

We started by creating an authenticated `requests` session by retrieving the Meraki API key from our environment and updating the header fields that are sent with each request. Please refer to the *How it works...* section of the *Storing authentication and metadata in requests sessions* recipe for a more detailed explanation of how this works.

With our session set up, we retrieved the QoS rules that were enabled. To do that, we pieced together a URL that included the ID of our network. Upon a successful response on our request to list all QoS rules, we used the ID of the QoS rule that is present in each of the QoS items in the list to issue a delete request for that specific rule, using both the network ID and the ID of the QoS rule. Then, we printed out an indication that the rule was successfully deleted if the `DELETE` request was successful. Notice how the success code for a `DELETE` request is usually `204` instead of `200`. Finally, we printed the status code for the user in case any of our requests failed.

There's more...

We can also use the Meraki SDK to achieve the same outcome, without having to manually do the requests ourselves. You must have the same environment variable (`MERAKI_DASHBOARD_API_KEY`) set:

```
import meraki

network_id = "<Insert network id here>"
dashboard = meraki.DashboardAPI()
```

```
rules = dashboard.switch.getNetworkSwitchQosRules(network_id)
for rule in rules:
    dashboard.switch.deleteNetworkQosRule(network_id,
rule['id'])
```

As you can see, the Meraki SDK takes care of putting together the correct URL based on our function call. It also takes care of sending and retrieving the request for us.

Using webhooks to programmatically react to an AP going down

How do you know when something goes wrong with your infrastructure? At the latest, this usually happens when a user complains about not being able to log into the network, but then, it is usually too late. The idea of monitoring your network for any kinds of unforeseen events, such as a rogue access point appearing within the premises of your network, is nothing new and with programmability, we could even automate parts of this. A simplistic approach would be to, every few seconds, send requests to our API and retrieve the information we need to make decisions regarding whether something is wrong or not. This approach, also referred to as *polling*, has multiple issues:

- We need to keep track of what information we have already seen and what information we have not seen yet.
- We are putting a lot of stress on the API by constantly sending requests.
- We are requesting a lot of information we don't really need.

Instead of polling, it would be easier if, any time something happens within our infrastructure, the API sends us a push notification stating that an event occurred. This is the idea of webhooks and in this recipe, we are going to use Flask, a lightweight framework for building web applications in Python, to receive these webhooks, check for a type, and print out an alert.

Getting ready

Open your code editor and start by creating a file called `webhook.py`. Next, navigate your terminal to the same directory that you just created the `webhook.py` file in.

While `requests` is great at *sending* HTTP requests, we now need to receive HTTP requests. To do this, we'll have to install a Python framework that allows us to build a lightweight web application. The framework is called Flask, and you can install it by running the `python3 -m pip install flask` command.

We will also need a HTTP tunnel since the Meraki cloud needs a way of sending a `POST` request to a server running on your local machine. You can use ngrok for this (`https://ngrok.com/`). After installation, you can start a new HTTP tunnel by opening a new terminal window and running the `ngrok http 8080` command. ngrok will display a forwarding address, as shown in the following screenshot:

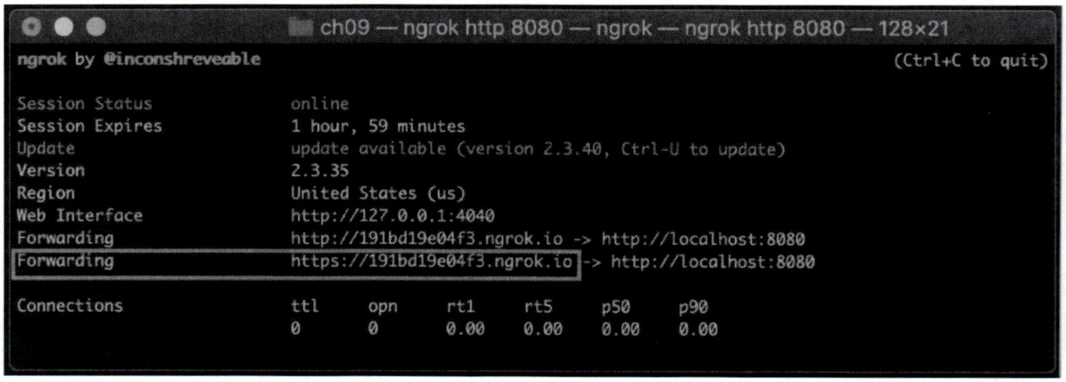

Figure 9.1 – ngrok displaying the forwarding address

You'll have to set up a webhook from your dashboard that points to the forwarding address specified by ngrok. You can find a guide on how to set up a webhook here: `https://documentation.meraki.com/General_Administration/Other_Topics/Webhooks`.

How to do it...

Follow these steps to create a lightweight web server that you can receive your Meraki webhooks on:

1. Import `flask`:

```
from flask import Flask, request
```

2. Create a new Flask `app`:

```
app = Flask("webhook_receiver")
```

3. In Flask, every HTTP request that gets sent to the server can be handled by a Python function. Set up a function for the root directory of our web server:

```
@app.route('/', methods=['POST'])
def webhook():
```

4. Within the webhook function, retrieve the data that is sent by the Meraki webhook:

```
        data = request.get_json()
```

5. Check if the alert type is from Air Marshal about a rogue AP and, if that's the case, print out information to the user:

```
        if data['alertType'] == "Air Marshal - Rogue AP
detected":
                print("Rogue AP detected!")
            return {'success': True}
```

6. Start the Flask app:

```
if __name__ == "__main__":
            app.run(host='0.0.0.0', port=8080)
```

7. Execute the script by running the following command in your terminal. Make sure that ngrok is still running in the other terminal window:

```
python3 webhook.py
```

You should see debug output that shows that your server is running. If a rogue AP is detected, you should see the line `Rogue AP detected` in the terminal.

How it works...

First, we imported the `flask` module and created a Flask app. This app abstracted away a lot of the complexity of actually receiving a web request. Instead, all we needed to do was provide a function. Every time someone hits the / endpoint on our little web server, this `webhook()` function will be executed. Within that function, we retrieved the data that was sent by the Meraki API. This data comes in JSON format, and we can use the `get_json()` function of the Flask request object to retrieve the data as a formatted Python dictionary. While every webhook can have different additional parameters, each of the webhooks has an `alertType` property that we can check. In this case, we only printed out a message to the command line in case a rogue AP came in. In reality, you could connect this to your chat application to automatically receive a notification or use services such as PagerDuty to page your on-call engineers.

With the handling done, we returned a dictionary with a success flag set to true. This triggers Flask to send a `200-OK` response back to the Meraki cloud. Lastly, we started our little web server on port `8080`, which is the same port that we had the ngrok tunnel forwarding all traffic to.

10
Incorporating your Python Scripts into an Existing Workflow by Writing Custom Ansible Modules

Ansible has, since its first release in early 2012, become a widely adopted tool for application deployment and infrastructure automation projects. Its vast ecosystem of pre-build modules allows you to describe the desired state of your infrastructure or application deployment using the **YAML Ain't Markup Language** (**YAML**) language. Being agentless, there is no need for a central controller that you send your instructions to for these to be carried out. Instead, your workstation running the description of tasks in the form of an Ansible playbook becomes the controller.

But have you ever wondered where all these modules that are pre-built for Ansible come from, and how they are implemented? In this chapter, we will get an introduction to developing your own Ansible modules. By developing modules, you can use Python to carry out complex workflows and then use Ansible and Ansible playbooks as an interface to run your workflows. In this chapter, we will cover the following recipes:

- Setting up the module structure
- Documenting your module
- Passing information into your module
- Using Ansible's built-in functionality to do web requests
- Packaging and calling your modules from Ansible playbooks

Technical requirements

For this section, you'll need a working installation of Ansible. Ansible is currently only supported on Unix or Unix-like systems such as most Linux distributions and Mac OS X. The code in this recipe was tested on both Linux and Mac OS X. If you are using Windows, you can either install Ansible in a Linux **virtual machine** (**VM**) or use the **Windows Subsystem for Linux** (**WSL**). You can find more information on how to set up Ansible under Windows by following this link to the official documentation: `https://docs.ansible.com/ansible/latest/user_guide/windows_faq.html#can-ansible-run-on-windows`.

You can install Ansible by running `python -m pip install --user ansible`, or if you are using Linux, by using the package manager of your distribution. You may find more information on how to set up Ansible in your specific environment by following this link to the official documentation: `https://docs.ansible.com/ansible/latest/installation_guide/intro_installation.html#installing-and-upgrading-ansible-with-pip`.

This section assumes that you have a working understanding of how Ansible functions from a user perspective, how to write playbooks, and how to run them.

You can view this chapter's code in action here: `https://bit.ly/3fW4m9o`

Setting up the module structure

Functionality in Ansible is provided and can be extended in two different ways: by writing either a **module** or a **plugin**. Ansible defines a module as "*reusable, standalone scripts that can be used by the Ansible API, the ansible command, or the ansible-playbook command. Modules provide a defined interface. Each module accepts arguments and returns information...*"(see `https://docs.ansible.com/ansible/latest/dev_guide/developing_locally.html#modules-and-plugins-what-is-the-difference`). An example of such a module could be one command that is executed on a remote device and retrieves some information for you. Writing your own modules can allow you to integrate internal tools, or tools where no module exists yet, into your existing Ansible-based automation plays. Modules for Ansible can be developed in many different languages, but since the core of Ansible is written in Python and most modules are also written in Python, we are going to focus on developing Ansible modules using Python. Plugins, on the other hand, are defined in the official documentation as "*Plugins are pieces of code that augment Ansible's core functionality.* "(see `https://docs.ansible.com/ansible/latest/plugins/plugins.html#plugins-lookup`). In this chapter, we are going to focus solely on modules.

In this recipe, we are going to see how to set up the directory and file structure necessary for writing your own modules, how to make Ansible aware of these new modules under development, and write and run our very first module.

Getting ready

Verify that Ansible is installed by opening up your terminal and running `ansible --version`.

How to do it...

Follow these steps to create your first Ansible module:

1. In your console, create a new directory that will contain the Ansible modules:

```
mkdir custom_ansible_modules
cd custom_ansible_modules
```

2. Next, we'll have to make Ansible aware of this new directory by adding it to the Ansible path. This path is controlled by the `ANSIBLE_LIBRARY` environment variable:

```
export ANSIBLE_LIBRARY=$ANSIBLE_LIBRARY:`pwd`
```

3. Verify that the path has been altered successfully by printing the `ANSIBLE_LIBRARY` path. The directory you are currently in should be at the end. If you do not see the path included, check the *There's more...* section of this recipe for possible solutions to this problem:

```
echo $ANSIBLE_LIBRARY
```

4. With our directory added to the path, we can create our first module by creating a `first_module.py` file:

```
touch first_module.py
```

5. With the structure created and Ansible made aware of our new module directory, you can open the previously created `first_module.py` file in your code editor to get started writing your first module.

   First, we need to import the required libraries from Ansible:

```
from ansible.module_utils.basic import AnsibleModule
```

6. Next, we want to define our `module` function in which we are going to specify the module itself:

```
def run_module():
```

7. Then, we'll define our module arguments as an empty dictionary:

```
args = {}
```

8. Next, we can define our `result` dictionary. The result is the data that is passed back from your module both to the user (by displaying it in your console) and to any subsequent modules that want to consume the information. In this simple example, we are going to pass a static method as well as information on the changed status:

```
result = {
            "changed": False,
            "message": "Hello from my ansible module"
}
```

9. Finally, we can define our Ansible module itself:

```
module = AnsibleModule(argument_spec=args,
                              supports_check_mode=False)
```

10. Since we only want to pass the static information specified in our `result` dictionary back, we can exit the module and return the content of the results as **JavaScript Object Notation (JSON)**:

```
module.exit_json(**result)
```

11. Finally, we need to specify that our previously defined `run_module()` method should be called when the file itself is being run:

```
if __name__ == "__main__":
        run_module()
```

12. With our first module written and stored in the previously created directory, we can go back to the terminal and run the newly created module:

```
ansible localhost -m first_module.py
```

The following screenshot shows what your output should look like:

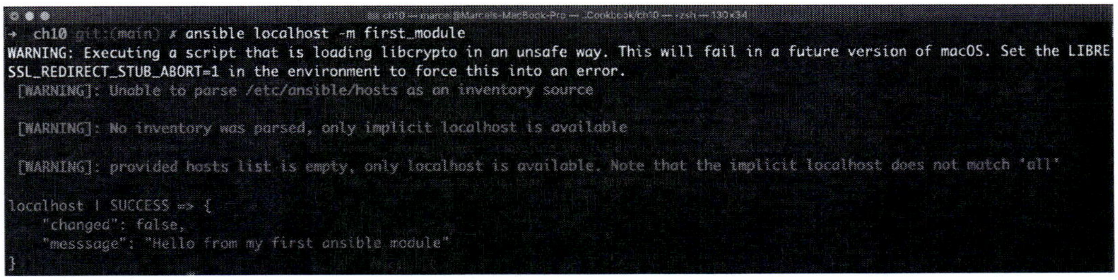

Figure 10.1 – Output from running our first Ansible module

How it works...

We start by setting up a directory that will contain all of the modules we want to develop. Next, we need to make Ansible aware of this new directory. By default, Ansible has some system-wide and user-specific directories such as `/usr/share/ansible/plugins/modules/` that will be searched when running a module from the command line or from a playbook. While we could put our newly created modules into one of those paths, it's easier for our development to define all our new modules in one central directory and then add that directory to the search path. We do this by changing the `ANSIBLE_LIBRARY` environment variable. This variable contains a list of additional paths Ansible should search in for modules. The list is separated by `:`.

After adding our new directory to the list of directories that are being searched by Ansible, we can create our first module, a Python file called `first_module.py`, in the previously created directory. This file then contains all the logic of our new module. We start by importing the required components from Ansible—namely, the `AnsibleModule` class. This class is the central structure of our module and is passed properties such as the arguments and returns so that they can then be used by Ansible. Next, we define a function called `run_module()`. This function will instantiate our new module. It is the function that is being run when our module is invoked by a playbook or by using the `ansible` command. We first specify a list of arguments. In our simple example, we don't accept any arguments, and thus `args` is just an empty dictionary. Next, we define our result. This result is the dictionary that is being returned back by our module. The information in this dictionary is returned to both any subsequent Ansible modules and the user in the form of a JSON-formatted string. Notably, there is a changed attribute that should be present. This Boolean attribute defines whether your code has changed the current status of a device or not. Modules that only gather information but don't *change* the configuration of the target device should always define `changed` as `false`.

With the inputs (the arguments in the `args` dictionary) and the outputs (the key-value pairs in our `result` dictionary) defined, we can create a module by instantiating `AnsibleModule`. The constructor takes two arguments, one being the previously defined list of available arguments, and the second, a flag that indicates that we do not support `check-mode`. Running a module in check-mode is like a dry run. Modules should not modify any information when being run in check-mode. Since we do not change any configuration in this module anyways, we can ignore check-mode for now. With the module defined, we can use its `exit_json(**results)` method to pass our previously defined results back to the system.

We then run our module against our localhost using the Ansible command-line tool, by specifying the name of our module without the `.py` extension. The output returned from Ansible is the dictionary that we passed back in the `result` variable of our Python script.

There's more...

Depending on your operating system and shell of choice, the backtick syntax that, in a shell script, executes the command that is written within the two backticks may not be supported. If you are seeing a `The module first_module was not found` error when trying to execute the module, this can be the issue. In that case, you have to manually set the correct library path:

1. Navigate to the directory in which you have put the `first_module.py` file. If you followed the instructions, this folder should be called `custom_ansible_modules`.

2. Inside of the `custom_ansible_modules` folder, print the absolute path to the folder:

```
pwd
```

3. Copy the path that is shown by the `pwd` command.
4. Add the absolute path to the `custom_ansible_module` folder to your `ANSIBLE_LIBRARY` environment variable:

```
export ANSIBLE_LIBRARY="$ANSIBLE_LIBRARY:<insert_the_
path_here>"
```

Documenting your module

One great feature of Ansible is its built-in documentation. Similar to man pages on a Unix system, we can use the `ansible-doc` command to get a summary of the module's functionality, which arguments it accepts, which values can be expected to be returned, as well as examples on how to use the module in your playbooks.

However, if we run the `ansible-doc` command on the module created in the *Setting up the module structure* recipe, you'll be left with an error message stating that the module is missing documentation:

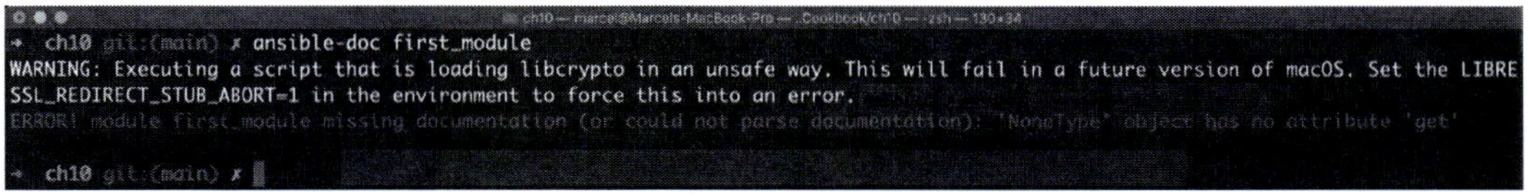

Figure 10.2 – Exemplar error when trying to access the documentation of an undocumented module

Ansible does not require us to provide any additional files to serve as the documentation that is retrieved and presented to the user by the `ansible-doc` command. Rather, we embed the documentation into the Ansible module itself, and this is exactly what we are going to do in this recipe.

Getting ready

Verify that Ansible is installed by opening up your terminal and running `ansible --version`, and also verify that you have included the local directory in which you are storing all your modules to the Ansible path by adding it to the `ANSIBLE_LIBRARY` environment variable. You can find information on how to do this in the *Setting up the module structure* recipe.

In that directory, create a new file called `documented_module.py` that is going to contain our module code.

How to do it...

Follow these steps to create a new Ansible module and specify the information necessary for `ansible-doc` to be able to show your module's documentation:

1. If you have not already done so, set up the directory structure needed for your Ansible modules. You can find a more detailed step-by-step description of this process in the *How to do it...* section of the *Setting up the module structure* recipe:

```
mkdir custom_ansible_modules
cd custom_ansible_modules
export ANSIBLE_LIBRARY=$ANSIBLE_LIBRARY:`pwd`
touch documented_module.py
```

2. With your directory structure set up, open the `documented_module.py` file in your code editor, and in it, start by importing the required modules from Ansible:

```
from ansible.module_utils.basic import AnsibleModule
```

3. We first specify the basic documentation information in a variable called `DOCUMENTATION`:

```
DOCUMENTATION = r'''
---
module: documented_module
short_description: A small documented module.
version_added: "1.0.0"
description: This is a longer description of our
documented module.
author:
            - Name (GitHub handle)
'''
```

4. With our basic information defined, we can also specify an example of how our basic module can be used in an Ansible playbook. We specify this information in a variable called `EXAMPLES`:

```
EXAMPLES = r'''
- name: Testing my documented task
        documented_module
'''
```

5. And finally, we can also specify the information that is being returned by our module by specifying a `RETURN` variable:

```
RETURN = r'''
message:
            description: A small welcome message returned
by our module
            type: str
            returned: always
            sample: 'Hello from my first ansible module'
'''
```

6. With our documentation defined, we can go ahead and actually specify the module and its functionality. Please refer to the *How to do it...* section of the *Setting up the module structure* recipe for a more detailed step-by-step description of how to specify your module:

```
def run_module():
    args = {}
    result = {
        "changed": False,
        "messsage": 'Hello from my first ansible module'
    }

    module = AnsibleModule(
        argument_spec=args,
        supports_check_mode=False
    )
    module.exit_json(**result)


def main():
    run_module()


if __name__ == '__main__':
    main()
```

7. With our module defined and documented, we can have a look at the docs by rerunning the `ansible-doc` command:

```
ansible-doc documented_module
```

The output returned by the `ansible-doc` command should look like this:

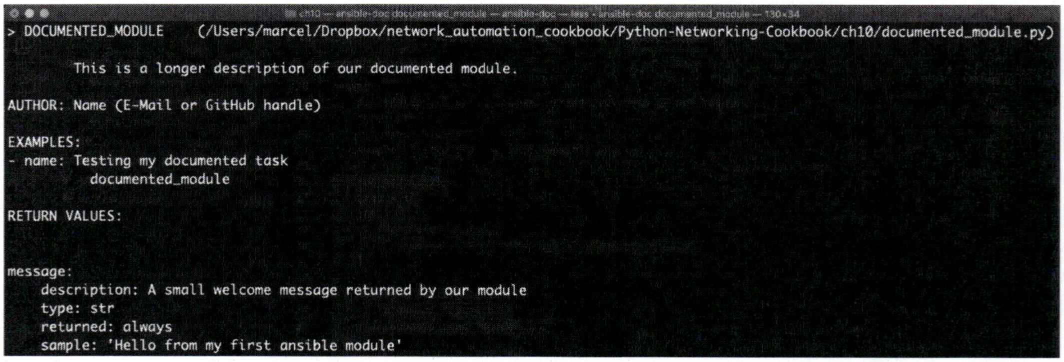

Figure 10.3 – Excerpt of the documentation for our module as it is shown in the ansible-doc command

How it works...

Ansible uses YAML not only for defining the playbooks but also for documenting its modules. Instead of having to provide an additional file that then contains this information, the YAML code is embedded directly into the source code of our module.

We first have to set up our directory structure for the new modules and make Ansible aware of where to search for them. For a more detailed description of this process, please refer to the *How it works...* section in the *Setting up the module structure* recipe.

With our module defined, we have three different variables: `DOCUMENTATION`, `EXAMPLES`, and `RETURN`. In order for the `ansible-doc` command to not error out, only the `DOCUMENTATION` variable is needed. Inside each of these variables, we define a raw multiline string that will contain our YAML code with the information required by `ansible-doc` to build the documentation. Using standard YAML syntax, we define the following:

- `module`, to contain the name of our module
- `short_description`, to contain a short summary of what our module is doing
- `version_added`, to specify the version of this module that was added to Ansible
- `description`, to contain a possibly longer description of what this module is doing
- `author`, to contain a list of the authors that have developed this module, with their name and GitHub handle

With the required information out of the way, we can then use the `EXAMPLES` variable to define one or more examples of how to use our module in a playbook. Additionally, we can use the `RETURN` variable to specify the return values that can be expected when running this module. The key (`message` in our example) that is being returned as part of the `result` dictionary then has the following properties that give more context around what is being returned:

- `description`, containing a description of what is being returned
- `type`, containing the data type that is being returned
- `returned`, containing information regarding the circumstances in which this property is returned; it could also be returned only on *success*
- `sample`, containing an exemplary return value

With our documentation specified, we can then go ahead and define the module itself. For a more detailed explanation of how this works, please refer to the *How it works...* section of the *Setting up the module structure* recipe.

Passing information into your module

So far, we have only dealt with a module that received no input from the user at all. One strength of Ansible is how easy a playbook makes it to structure the inputs that you want to pass from a module, as well as passing the output of one module back into the input of another. In this recipe, we are going to see how to both implement and document the arguments that are being passed into our module.

Getting ready

Verify that Ansible is installed by opening up your terminal and running `ansible --version`, and also verify that you have included the local directory in which you are storing all your modules to the Ansible path by adding it to the `ANSIBLE_LIBRARY` environment variable. You can find information on how to do this in the *Setting up the module structure* recipe.

In that directory, create a new file called `value_module.py` that is going to contain our module code.

How to do it...

Follow these steps to create a new Ansible module and specify the information necessary for `ansible-doc` to be able to show your module's documentation:

1. If you have not already done so, set up the directory structure needed for your Ansible modules. You can find a more detailed step-by-step description of this process in the *How to do it...* section of the *Setting up the module structure* recipe:

```
mkdir custom_ansible_modules
cd custom_ansible_modules
export ANSIBLE_LIBRARY=$ANSIBLE_LIBRARY:`pwd`
touch value_module.py
```

2. With your directory structure set up, open the `value_module.py` file in your code editor, and in it, start by importing the required modules from Ansible. Additionally, we will import the built-in `ipaddress` module to generate our **Internet Protocol** (**IP**) addresses:

```
from ansible.module_utils.basic import AnsibleModule
import ipaddress
```

3. We first specify the basic documentation information in a variable called `DOCUMENTATION`:

```
DOCUMENTATION = r'''
---
module: value_module
short_description: A small module to create IPv4 address
list.
version_added: "1.0.0."
description: Create a IPv4 address list based on host
portion and netmask information.

options:
    ip_address:
        description: The network portion of your IP
        required: true
        type: str
    netmask:
        description: The netmask of your IP address pool
```

```
            required: true
            type: int

author:
    - Name (GitHub handle)
'''
```

4. With our basic documentation done, we can also document the results and examples:

```
EXAMPLES = r'''
- name: Testing my ip address task
      value_module:
        ip_address: 192.168.10.0
        netmask: 28
'''

RETURN = r'''
addressses:
    description: A list of IPv4 addresses
    type: list
    returned: always
    sample: ['192.168.10.1', '192.168.10.2']
'''
```

5. With our documentation done, we can start specifying the module itself inside our `run_module()` method:

```
def run_module():
```

6. We start by defining two parameters inside of the `args` dictionary:

```
    args = {
        "ip_address": {
            "type": str,
            "required": True
        },
        "netmask": {
```

```
            "type": int,
            "required": True
        }
    }
```

7. Next, we define the module object itself and pass the arguments specified in the preceding step:

```
    module = AnsibleModule(
        argument_spec=args,
        supports_check_mode=False
    )
```

8. Next comes our logic of actually generating the IP addresses using the `ipaddress` module and the parameters provided by the Ansible playbook or command line:

```
    # Generate IP addresses

         cidr_address = f"{module.params['ip_address']}/
  {module.params['netmask']}"

    address_objects = ipaddress.IPv4Network(cidr_address)

    addresses = []
    for a in address_objects:
        addresses.append(str(a))
```

9. We then return the information created in the preceding snippet as the result of our module:

```
    result = {
        "changed": False,
        "addresses": addresses
    }
    module.exit_json(**result)
```

10. Finally, we need to instruct Python to run the `run_module()` function:

```
if __name__ == '__main__':
    run_module()
```

11. We can test the functionality of our module by running the following command:

```
ansible localhost -m value_module -a "ip_
address=10.10.10.0 netmask=28"
```

Your output should look like this:

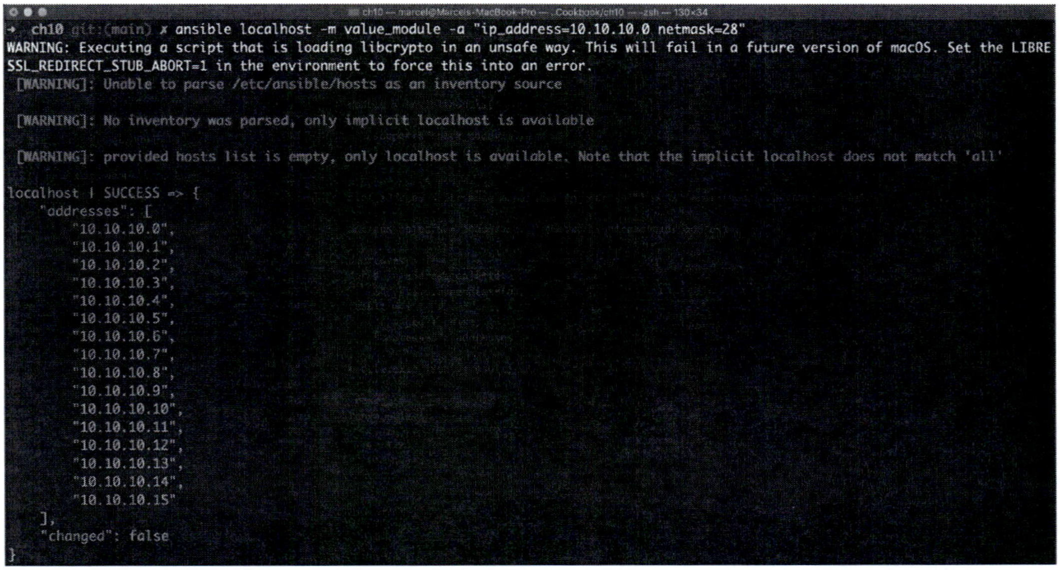

Figure 10.4 – Output from our module showing the IP pool generated by our module

How it works...

We first have to set up our directory structure for the new modules and make Ansible aware of where to search for them. For a more detailed description of this process, please refer to the *How it works...* section of the *Setting up the module structure* recipe.

With our directory structure set up, we can import Ansible and the module required to carry out our operations before defining the documentation of our module. Please refer to the *How it works...* section of the *Documenting your module* recipe for a description of the different documentation variables. In addition to the information covered in that section, we now also include a summary of our input arguments. In this case, we are documenting two different variables, an IP address of type `str` and a netmask of type `int`.

With our documentation done, we can go ahead and specify our actual module. We first define the arguments that we require in the `args` dictionary. For each argument, we specify the name, data type, and whether it is a required argument or not. Ansible will make sure that these parameters are then present when the module is run or will return an error to the user, stating which variables are missing.

With the arguments dictionary defined, we then instantiate our module before generating the IP addresses. Notice how we use the `params` variable of our module to access the information passed into the module. We use that information to piece together a **classless inter-domain routing** (**CIDR**) string on which we can use the `ipaddress` module to generate a list of IP addresses. We convert these into strings since we are ultimately returning JSON, and thus every variable in the `result` dictionary needs to be JSON-serializable.

We then call our module and, using the `-a` command-line parameter, are able to pass the arguments required by the module.

Using Ansible's built-in functionality to do web requests

In previous chapters, we have seen how to use modules from the Python standard library within our Ansible module. But what about third-party packages such as the `requests` package we used in *Chapter 8, Configuring Devices Using RESTCONF and requests*, and *Chapter 9, Consuming Controllers and High-Level Networking APIs with requests*, to make web requests? While we could import an external package, Ansible modules generally try to be as independent of third-party packages as possible. But this does not mean that we have to rebuild the entire functionality to make **HyperText Transfer Protocol** (**HTTP**) requests from scratch. Ansible comes pre-packaged with many built-in utility modules, one of which, the `uri` module, can be used to make web requests.

In this recipe, we will see how you can use an Ansible module and built-in functionality to send an HTTP request to the Meraki **application programming interface** (**API**). We'll be using the Meraki API as an example here, and the same code would work for different APIs. For an introduction to consuming **REpresentational State Transfer** (**REST**) APIs in Python, please refer to *Chapter 9, Consuming Controllers and High-Level Networking APIs with requests.*

Getting ready

Verify that Ansible is installed by opening up your terminal and running `ansible --version`, and also verify that you have included the local directory in which you are storing all your modules to the Ansible path by adding it to the `ANSIBLE_LIBRARY` environment variable. You can find information on how to do this in the *Setting up the module structure* recipe.

In that directory, create a new file called `web_module.py` that is going to contain our module code.

How to do it...

Follow these steps to create a new Ansible module and specify the information necessary for `ansible-doc` to be able to show your module's documentation:

1. If you have not already done so, set up the directory structure needed for your Ansible modules. You can find a more detailed step-by-step description of this process in the *How to do it…* section of the *Setting up the module structure* recipe:

```
mkdir custom_ansible_modules
cd custom_ansible_modules
export ANSIBLE_LIBRARY=$ANSIBLE_LIBRARY:`pwd`
touch web_module.py
```

2. With your directory structure set up, open the `web_module.py` file in your code editor, and in it, start by importing the required modules from Ansible. Additionally, we will import the build-in JSON module to parse our API output, as well as a built-in helper function to make web requests:

```
from ansible.module_utils.basic import AnsibleModule
form ansible.module_utils.urls import fetch_url
import json
```

3. We first specify the basic documentation information in a `DOCUMENTATION` variable:

```
DOCUMENTATION = r'''
---
module: web_module
short_description: A small module to retrieve all orgs
from the meraki API.
version_added: "1.0.0."
description: Retrieve all organizations the user has
access to.

options:
    access_token:
        description: The API access token to use
        required: true
        type: str
    base_url:
```

```
        description: The base url of the meraki API to use
        required: false
        default: https://api.meraki.com/api/v1
        type: str

author:
    - Name (GitHub handle)
'''
```

4. With our basic documentation done, we can also document the results and examples:

```
EXAMPLES = r'''
- name: Retrieving all my organizations from meraki
       web_module:
        access_token: <Insert access token here>
'''

RETURN = r'''
status:
    description: The returned http status
    type: int
    returned: always
    sample: 200
organizations:
    description: A list of dictionaries describing the
organizations
    type: list
    returned: success
    sample: [{"id": "463308", "name": "DevNet San
Jose", "url": "https://n18.meraki.com/o/vB2D8a/manage/
organization/overview"},]
msg:
          description: A status message in case of an
unsuccesful request
          type: str
          returned: failure
          sample: Unable to call API.
'''
```

5. With the documentation finished, we can start specifying the module itself inside our `run_module()` method:

```
def run_module():
```

6. We start by defining two parameters inside of the `args` dictionary:

```
    args = {
        "access_token": {
            "required": True,
            "type": str
        },
        "base_url": {
            "required": False,
            "default": "https://api.meraki.com/api/v1",
            "type": str
        }
    }
```

7. Next, we can create our module:

```
    module = AnsibleModule(
        argument_spec=args,
        supports_check_mode=False
    )
```

8. For our web request, we will need headers that contain the authentication information (the access token passed into the module as a parameter). We can set them up as a dictionary:

```
    headers = {
        "X-Cisco-Meraki-API-Key":module.params['access_
token']
    }
```

9. Next, we can piece together our request **Uniform Resource Locator** (**URL**) and send the actual request:

```
    url = "{}/organizations".format(module.params['base_
url'])
    resp, info = fetch_url(module, url, method="get",
headers=headers)
```

10. We then need to prepare our resulting output:

```
    result = {
        "changed": False,
        "status": info['status'],
    }
```

11. In case of a successful request, we want to return all the organizations. In case of an unsuccessful return from the API, we want to return an error message together with the status code:

```
    if info['status'] == 200:
        result["networks"] = json.loads(resp.read())
        module.exit_json(**result)
    else:
        result["msg"] = "Unable to call API."
        module.fail_json(**result)
```

12. Finally, we need to instruct Python to run the `run_module()` function:

```
if __name__ == '__main__':
    run_module()
```

13. We can test the functionality of our module by running the following command:

```
ansible localhost -m web_module -a "access_
token=6bec40cf957de430a6f1f2baa056b99a4fac9ea0"
```

Your output should look like this:

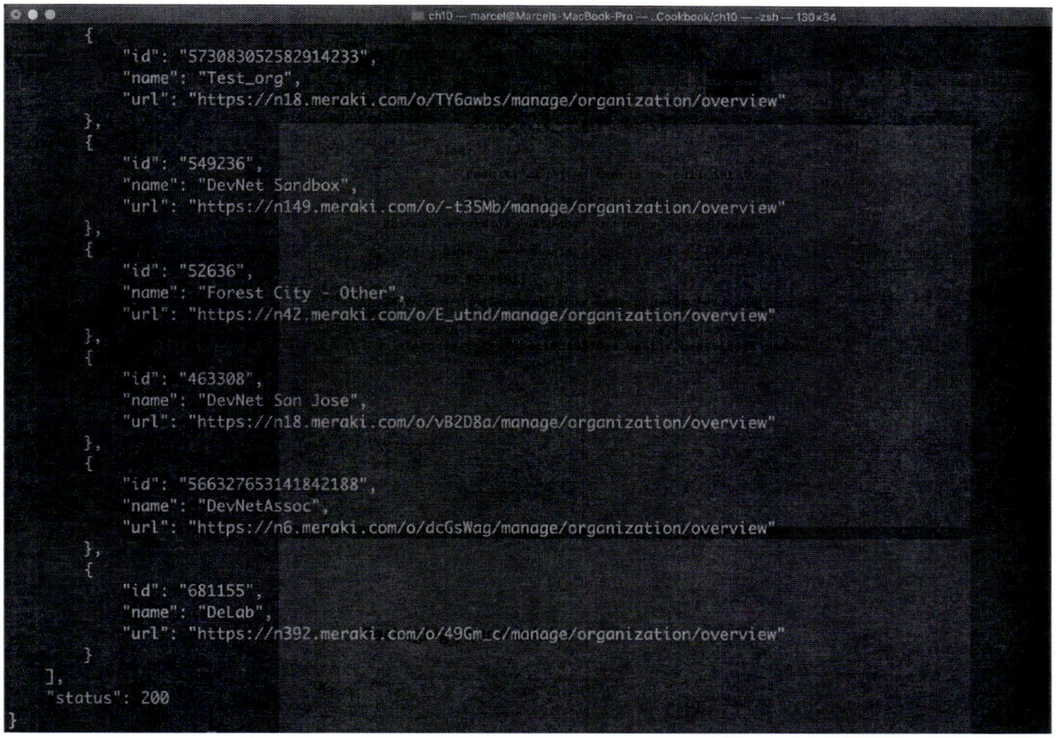

Figure 10.5 – Exemplar output from querying all Meraki organizations in our account using our module

The access token mentioned here is for the public read-only Cisco Meraki sandbox, and you can use it to test your module.

How it works...

We first have to set up our directory structure for the new modules and make Ansible aware of where to search for them. For a more detailed description of this process, please refer to the *How it works...* section in the *Setting up the module structure* recipe.

With our directory structure set up, we can import Ansible and the JSON module from the standard library. Additionally, we import the `fetch_url` method from Ansible's built-in module of helper functions around the querying of web services. Next, we document our new module. Please refer to the *How it works...* section of the *Documenting your module* recipe for a description of the different documentation variables. Additionally, we also see an optional parameter, the base URL, indicated by a `required: false` definition. This optional parameter also has a default value specified that is used when the flag is not passed.

With our documentation done, we can then start implementing the module functionality itself inside the `run_module()` function. We start by specifying our input arguments. This time, we not only create a required flag in the form of our access token but also an optional flag in the form of our base URL. It's in the `args` dictionary where we specify the default value that Ansible will use if this parameter is not set by the user in their playbook that's calling the module.

With the object of our Ansible module initialized, we can then go ahead and carry out our actual workflow—the web request to the Meraki API. To do this, we first specify headers. Similar to how requests use dictionaries to represent their headers, we can also do the same here and, as such, we specify a single additional header in the form of our `X-Cisco-Meraki-API-Key` header, whose value we retrieve from the parameters passed by the playbook or your command line.

Next, we do the actual request itself by first piecing together the target URL. To do so, we retrieve the base URL from our parameters. If the user did not specify any overwriting base URL argument, it will be the same value as the default specified in the `args` dictionary. We then use the `fetch_url()` function to run the web request itself. Finally, we set up our `result` dictionary. Irrespective of failure or success, we always want to return the fact that nothing changed (`changed=False`) and the status code of our web request, which we can retrieve from the `info` dictionary returned by the `fetch_url()` function. We then differentiate a successful request and an unsuccessful request based on the returned status code and, depending on whether we were successful or not, we return a different `result` dictionary. In the case of success, our `result` dictionary uses the JSON module from the standard library to parse the body of the web response, and thus returns a list of all organizations this user has access to. In the case of an unsuccessful request, a message stating that the request was unsuccessful is returned.

We then call our module and, using the `-a` command-line parameter, are able to pass the arguments required by the module.

Packaging and calling your modules from Ansible playbooks

In the recipes so far, we have always used the Ansible command-line tool to directly invoke a module. While this is great for debugging and testing during our development cycle, the goal of Ansible modules is usually to be part of an Ansible playbook. So, how can we use our newly created Ansible modules from a playbook?

In this recipe, we will create a simple playbook as well as the folder structure necessary to call the `web_module` module that we built in the *Using Ansible's built-in functionality to do web requests* recipe. The playbook will call our self-written module and then use an Ansible `debug` task to print out the name of all our organizations that we have access to.

Getting ready

Verify that Ansible is installed by opening up your terminal and running `ansible --version`. You will need a version of our previously written `web_module.py` file. You can obtain this by either following the instructions in the *Using Ansible's built-in functionality to do web requests* recipe or by downloading it from the book's GitHub repository. You can find the files for this chapter by accessing this link: `https://github.com/PacktPublishing/Python-Networking-Cookbook/tree/main/ch10`.

Create a new directory called `module_test`. Inside this directory, you'll create a file called `sample_playbook.yaml` and another directory called `libraries`.

How to do it...

1. Copy the previously created or obtained `web_module.py` file into the `libraries` folder.
2. Open up the `sample_playbook.yaml` file. Inside this file, we'll write our playbook.
3. We first need to specify the host; in our case, we are running this on our localhost:

```
- hosts: localhost
```

4. Next, we specify a list of tasks. Our first task will be to call our custom module to retrieve the organizations:

```
tasks:
    - name: Testing our module from ansible playbook
      web_module:
        access_token:
6bec40cf957de430a6f1f2baa056b99a4fac9ea0
```

5. In that task, we also want to make sure that we register the output as a new variable called `result` and delegate this to the localhost so that it runs on our local machine:

```
      register: result
      delegate_to: localhost
```

6. With the list of networks retrieved and stored in the `result` variable, we can use an Ansible `debug` task to then retrieve the list of names:

```
    - name: Show results
      debug:
```

7. Inside of the `debug` task, we iterate over each item in the organization list that is contained within our `result` object and print the name:

```
        msg: "{{ item.name }}"
      with_items: "{{ result['networks'] }}"
```

8. With our playbook specified, you can now run it by using the following command:

```
ansible-playbook sample_playbook.yaml
```

Your output should look like this:

Figure 10.6 – Exemplar output from calling our self-developed module from an Ansible playbook

How it works...

By placing the previously created module in a `libraries` folder that is in the same parent directory as our playbook itself, our self-developed module is made available to the playbook when we are running it. We can thus use our self-developed module in the same way as any module included with Ansible or installed from Ansible Galaxy. In our playbook, we first define that we only want to run this on `localhost` before specifying a list of tasks. Our first task is to run our `web_module` module. Similar to how we passed the access token as a parameter using the `-a` command-line option when running the module previously, we now have to pass it from our playbook. We then register the output of our module to be saved in the `result` variable.

With our module called, we can then print out the output. We use the `debug` task that comes with Ansible, together with the built-in `with_items` directive, to iterate over all items in our list of networks. Remember that, as with the other modules, our custom module just returned JSON, and Ansible can thus easily parse the module output and make it available for simple looping operations. In our case, we loop over all organizations and then print out the name as the debug message.

There's more...

You might encounter that your version of Ansible is displaying the name of the task with a little drawing of a cow and a speech bubble:

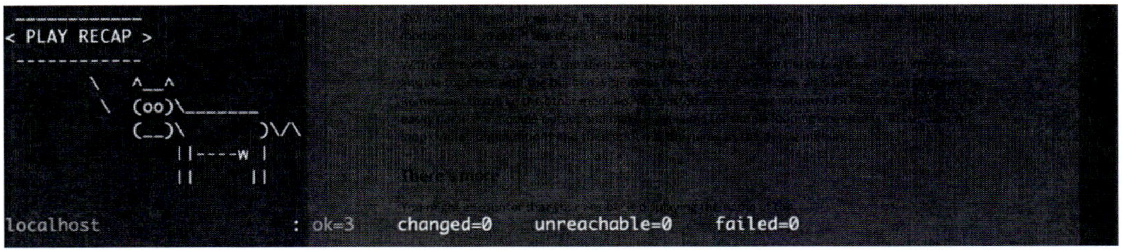

Figure 10.7 – Ansible's default behavior of using cowsay American Standard Code for Information Interchange (ASCII) art to display debug messages

You can disable this behavior by setting the `ANSIBLE_NOCOWS=1` environment variable.

11
Automating AWS Cloud Networking Infrastructure Using the AWS Python SDK

Very few new technology trends had such a big impact on infrastructure in the past years as cloud computing. Spearheaded by the demand for computing infrastructure that is available on-demand and billed to usage, cloud computing providers such as Amazon Web Services, Google Cloud, and Microsoft Azure have influenced how people think about infrastructure as a whole. Gone are the days where you had to have a data center with racks upon racks of servers to deploy your application to. Today, you can farm out many of the administrative tasks of running a data center to your cloud providers and just consume the resources you need. One key benefit of this flexibility is that it perfectly matches the concepts of **Infrastructure as Code** (**IaC**) and programmability. Using the cloud provider's APIs, we can automatically create, administer, and destroy resources such as the computing infrastructure in the cloud, and this is exactly what we will look at in this chapter while using **Amazon Web Services** (**AWS**).

While we could use the techniques we explained in *Chapter 9, Consuming Controllers and High-Level Networking APIs with requests*, and leverage the REST APIs provided by AWS, we are going to use the boto3 package in this chapter. With boto3, AWS has developed a **software development kit** (**SDK**) that abstracts the REST API for us as developers and lets us carry out operations in a more pythonic way.

In this chapter, we will be covering the following recipes:

- Setting up the library to interact with your AWS account
- Collecting information about your cloud networking resources
- Starting an EC2 instance
- Creating a VPC
- Subnetting your VPC
- Changing routes in your VPC

Technical requirements

For this chapter and the remainder of this book, you'll need to install Python. Specifically, you'll need a Python interpreter that's version 3.6.1 or higher. This book makes use of the language constructs provided by Python 3, which means it is incompatible with Python 2.x. Additionally, you'll need to install the boto3 package. You can install the newest version of boto3 using `python3 -m pip install boto3`. At the time of writing, the current version is version 1.17.91.

You will also need a code editor. Popular choices include Microsoft Visual Studio Code and Notepad++. Additionally, you'll need an AWS account. You can sign up for a free tier account that will give you some limited computing resources to try out the scripts in this chapter.

A basic understanding of the different concepts and services, such as EC2, is advisable when following the recipes in this chapter. You may gather such an overview of key concepts by visiting the *Getting started* page from AWS at `https://aws.amazon.com/getting-started/`.

You can view this chapter's code in action here: `https://bit.ly/3fXAObg`

Setting up the library to interact with your AWS account

When logging into the AWS web console, you are probably used to using a username/password combination to authenticate either for an organization's root user account, or for an IAM user that has been set up with a separate set of credentials and scoped access. When dealing with AWS programmatically using the boto3 SDK, we will use two different credentials, an **Access Key ID** and a **Secret Key**, to authenticate against the API. In this recipe, we are going to configure our key pair in the web console, which we'll then use for the remainder of this chapter to authenticate our requests. While there are multiple ways to pass this credential information to boto3, we are going to configure it in a central configuration file for convenient reuse.

Getting ready

You'll need an AWS account, either for a root user or for an IAM account that has been created with the necessary access scopes for you.

How to do it...

Follow these steps to create the credentials you are going to use for authenticating your requests from the SDK:

1. Open the AWS console by accessing `console.aws.amazon.com` and authenticating using your **username/password** combination.
2. Navigate to your user's security credentials:

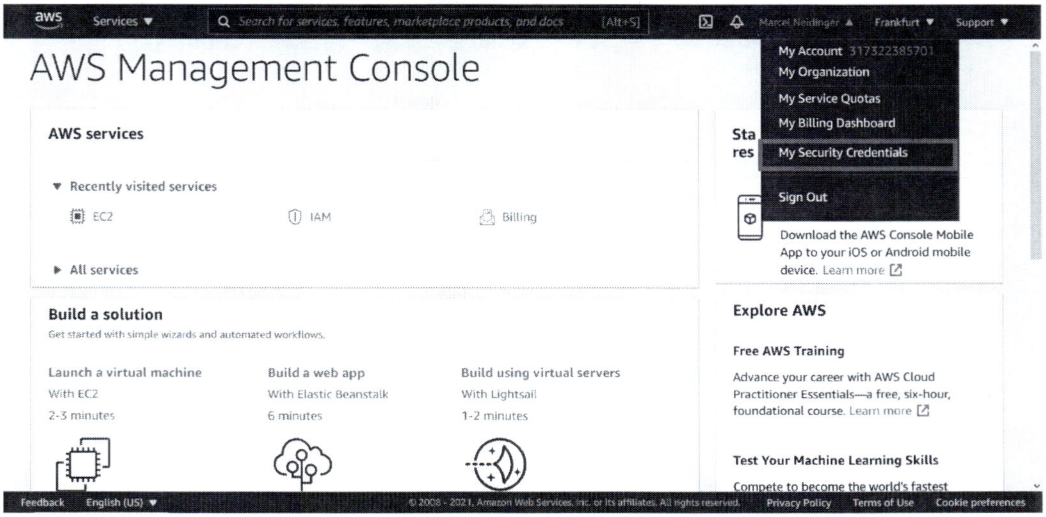

Figure 11.1 – Start page of the AWS Management Console with the credentials settings highlighted

3. On the Your **Security Credentials** page, navigate to the **Access keys** section and click on **Create New Access Key** to generate a new key pair:

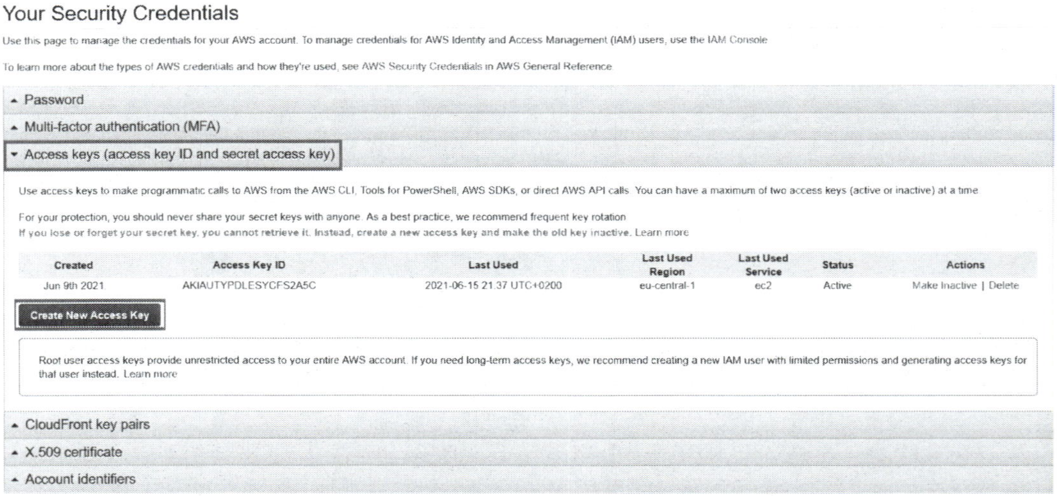

Figure 11.2 – The "Your Security Credentials" section, where you can generate a new pair of access keys

4. You'll be presented with a pop-up that shows your newly created Access Key ID and Secret Access Key. You can either save them by copying the information from the page or by clicking the **Download Key File** button. Take note of your **Secret Access Key**. Once this pop-up window has been closed, you won't be able to retrieve this key again:

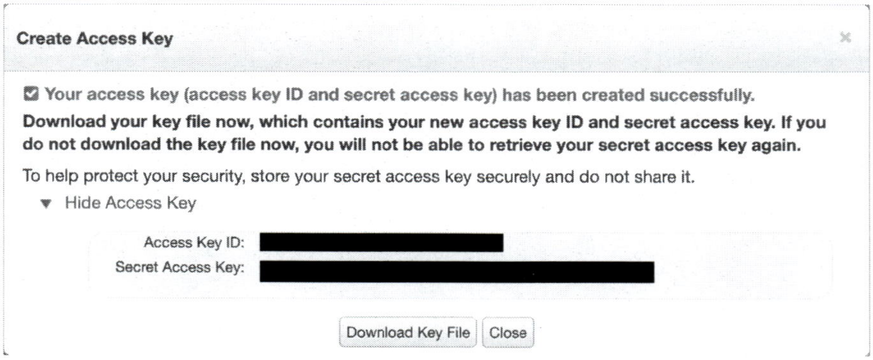

Figure 11.3 – Example pop-up with the newly created credentials

5. Create a credentials file by opening your text editor and creating the file in the `~/.aws/credentials` path. Then, add your credentials to it:

```
[default]
aws_access_key_id = <Insert your access key here>
aws_secret_access_key = <Insert your secret key here>
```

6. You will also want to configure your default region. To do so, create a file in the `~/.aws/config` path and add the following configuration. Have a look at the *How it works...* section for a list of available regions:

```
[default]
region = <insert your region here, i.e. us-east-1>
```

How it works...

In this recipe, we used the AWS console to generate a new key pair to use when programmatically accessing AWS resources. With the retrieved credentials, we set up our configuration so that boto3 knows where to find them. We added our previously created Access Key ID and Secret Key to an INI-style credentials file and then configured the default region. The following is a (non-comprehensive) list of the available regions:

- `us-east-1`
- `us-west-1`
- `eu-central-1`
- `eu-west-1`
- `ap-east-1`
- `ap-south-1`
- `me-south-1`

You can find a full list of all available regions at `https://docs.aws.amazon.com/AmazonRDS/latest/UserGuide/Concepts.RegionsAndAvailabilityZones.html`. With the credentials set, the boto3 library will be able to pick them up without us having to specifically provide the information in the code.

There's more...

In this recipe, we created a central configuration file that will be automatically used by boto3. Especially when using your script in a **continuous integration/continuous delivery** (**CI/CD**) pipeline, you may not want to create such a configuration on the server running your deployment script. As an alternative, you can also specify them using environment variables. These environment variables are as follows:

- `AWS_ACCESS_KEY_ID` for your AWS Access Key ID
- `AWS_SECRET_ACCESS_KEY` for your AWS Secret Access Key
- `AWS_DEFAULT_REGION` for your default region

The values you must provide here are the same as the ones you inserted into the config files in this recipe.

If you have the AWS CLI tools installed, you can also run the following command, which will guide you through the process of creating the configuration files:

```
aws configure
```

Collecting information about your cloud networking resources

With our programmatic connection to AWS set up, we can now start using the boto3 library to gather some facts about the deployments that are currently running in the account. This is a common task that can come in useful when we are trying to answer questions about how many **Virtual Private Clouds** (**VPCs**) we have, or what **Elastic Compute Cloud** (**EC2**) instance is associated with which private cloud. While all this information is also available in the graphical user interface provided by the AWS console, being able to retrieve such information programmatically can then enable us to automate maintenance tasks such as deleting VPCs that do not have any EC2 instances associated with them. In this recipe, you are going to learn how to programmatically retrieve all the VPCs that are associated with your account and, for each of these VPCs, retrieve the EC2 instances that are currently part of that specific VPC. AWS and the boto3 library use the same structure to deal with requests for all their resources, which means you can easily adapt this script to filter for different parameters than the ones we are going to see here.

Getting ready

Open your code editor and start by creating a file called `get_facts.py`. Next, navigate your terminal to the same directory that you just created the `get_facts.py` file in.

You'll have to have authentication set up, as described in the *Setting up the library to interact with your AWS account* recipe.

You will also need a running EC2 instance that you can launch from the console. A `t2.micro` instance that is eligible for the free tier is sufficient here. AWS will create a default VPC for you when you create your EC2 instance.

How to do it...

Follow these steps to programmatically retrieve all the VPCs and their associated EC2 instances:

1. Import the `boto3` library:

```
import boto3
```

2. Set up the `ec2` client from `boto3`:

```
ec2 = boto3.client('ec2')
```

3. First, retrieve all the VPCs:

```
vpc_data = ec2.describe_vpcs()
```

4. Then, iterate over all the VPCs:

```
for v in vpc_data['Vpcs']:
    print(f"VPC Id: {v['VpcId']}")
```

5. Filter for all EC2 instances associated with this VPC by providing a `Filters` argument:

```
    instance_data = ec2.describe_instances(Filters=[
        {
            "Name": "vpc-id",
            "Values": [
                v['VpcId']
            ]
        }
    ])
```

6. Check if we have active reservations:

```
    if len(instance_data['Reservations']) > 0:
```

7. Then, iterate over all the active `Reservations` groups:

```
        for res_num in range(0, len(instance_
data['Reservations'])):
```

8. Finally, retrieve all the instances in that reservation group and display some information for that instance:

```
        for i in instance_data['Reservations'][res_
num].get('Instances', []):
            print(f"- {i['InstanceType']} launched at
{i['LaunchTime']}")
```

9. Run your script by typing the following command in your console:

```
python3 get_facts.py
```

The output should look as follows:

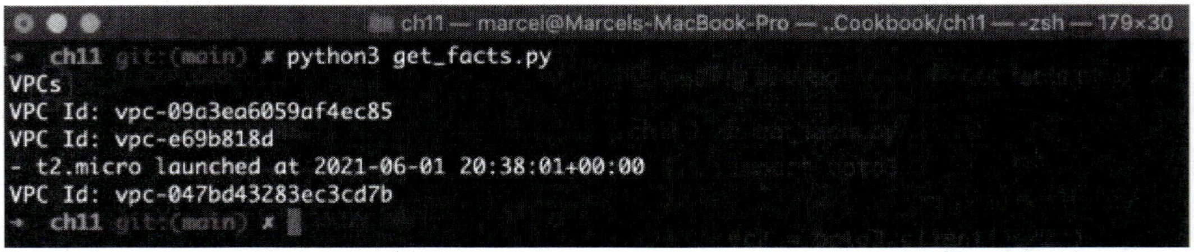

Figure 11.4 – Exemplary output from our facts retrieval script. The number of VPCs and IDs will be different

How it works...

We started by importing the `boto3` library, our SDK for the AWS REST API. Within `boto3`, every resource, such as the S3 storage or the EC2 compute part, has its own client that we can retrieve using the service identifier, which in this case is `ec2`. The returned client object has a long list of available methods that we can use to retrieve, create, manipulate, or delete resources that are part of this section. These functions follow a naming convention:

- Functions starting with `retrieve_` allow you to retrieve one or more instances of the resource, as well as filter on them.
- Functions starting with `delete_` allow you to delete resources.
- Functions starting with `create_` allow you to create resources.
- Functions starting with `modify_` allow you to modify resources.

A full list of available functions (at the time of writing, there are 453 just for the EC2 client) can be found in the following documentation: `https://boto3.amazonaws.com/v1/documentation/api/latest/reference/services/ec2.html`.

Using the `describe` function, we retrieved all the available VPCs. In return, we received a list of dictionaries that contain all the information about our VPCs. The following screenshot of the documentation shows the response's structure:

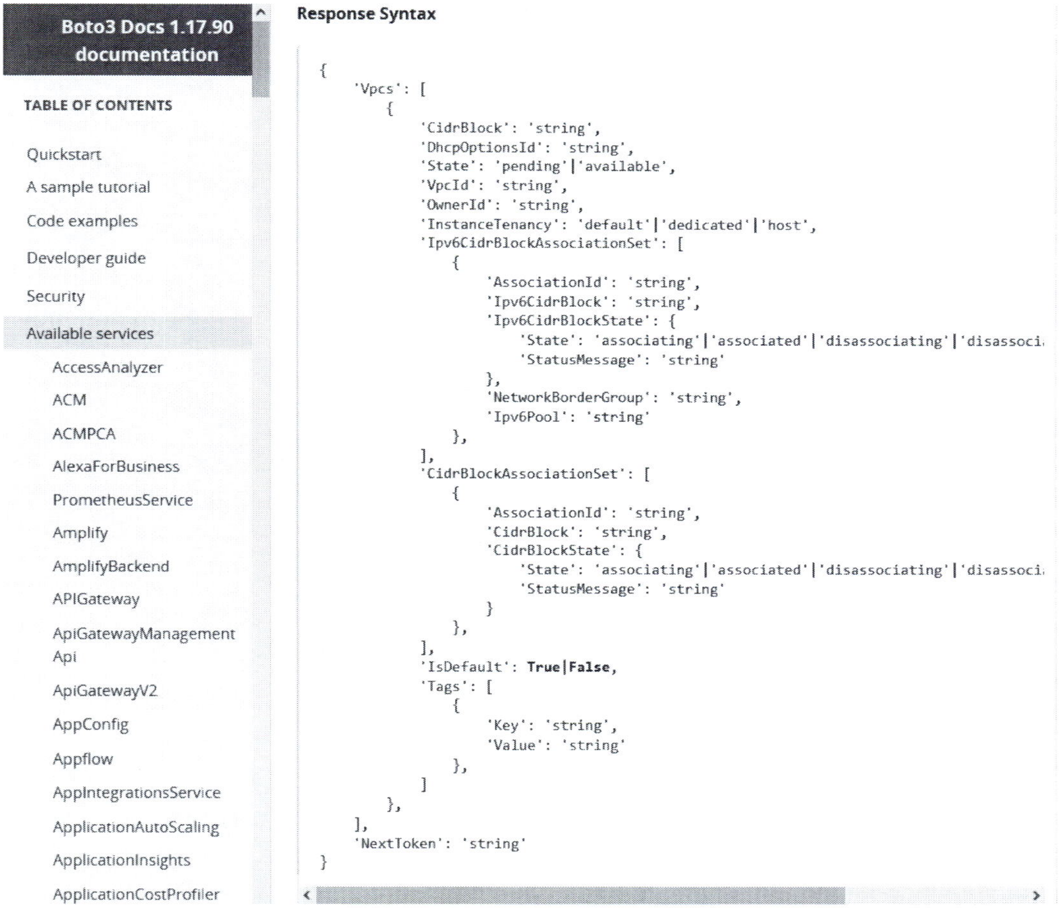

Figure 11.5 – Response structure of a VPC

Below the structure, there is also a table containing the names and descriptions of the returned keys. You can find this list in the following documentation: `https://boto3.amazonaws.com/v1/documentation/api/latest/reference/services/ec2.html#EC2.Client.describe_vpcs`.

With our VPC ID, we filtered all our EC2 instances for those running within the provided `Vpc Id`. The `describe_instances()` function takes a `Filter` argument. Filters are always a list of dictionaries. These dictionaries must have the same keys:

- A `Name` key that specifies the name of the filter.
- A `Values` key that specifies a list of values to filter for. This always needs to be a list, even if we are, as in this example, filtering for only one value (our `Vpc Id`).

A list of all the filters is available in the following documentation: `https://boto3.amazonaws.com/v1/documentation/api/latest/reference/services/ec2.html#EC2.Client.describe_instances`.

With the filters specified, we iterated over all the instances and printed out some basic information, such as the instance type and when this instance was launched. The response syntax is also referenced in the documentation.

Starting EC2 instances

In the previous recipes, we only retrieved facts about our currently running infrastructure. But we still had to create the necessary resources so that something shows up using the console. While this is a painless process in the web browser, in this recipe, we want to explore how to launch an EC2 instance using boto3.

Getting ready

Open your code editor and start by creating a file called `create_ec2.py`. Next, navigate your terminal to the same directory that you just created the `create_ec2.py` file in.

You'll have to have the authentication set up as described in the *Setting up the library to interact with your AWS account* recipe.

Additionally, you will need a previously set up SSH key pair. This is not the key pair that we used for authenticating to the API, but rather an SSH key that can be used to log into the instance itself via SSH once launched. You can create this key pair in the console. You can find a guide on how to do this at `https://docs.aws.amazon.com/AWSEC2/latest/UserGuide/ec2-key-pairs.html`.

How to do it...

Follow these steps to launch an EC2 instance from your Python script:

1. Import the `boto3` library:

```
import boto3
```

2. Create an `ec2` resource:

```
ec2 = boto3.resource('ec2')
```

3. Next, we need to specify what kind of instance (the image) we would like to launch and what `key-pair` can be used for SSH authentication. Have a look at the *How it works...* section for a (non-comprehensive) list of available images:

```
KEY_PAIR = "<Insert key-pair name here>"
IMAGE_ID = "<Insert image id here>"
NUM_INSTANCES = <Insert number of instances>
```

4. With our information specified, we can create our new instance:

```
i = ec2.create_instances(ImageId=IMAGE_ID,
                         KeyName=KEY_PAIR,
                         MaxCount=NUM_INSTANCES,
                         MinCount=NUM_INSTANCES)
```

5. Finally, print out the instance:

```
print(i)
```

6. Run the script by typing the following command:

```
python3 create_ec2.py
```

The output of your script should look similar to the following. The ID of your instance will be different:

Figure 11.6 – Screenshot showing the desired output of running our script

You can verify that the instance has been created by accessing the list of instances in your AWS console:

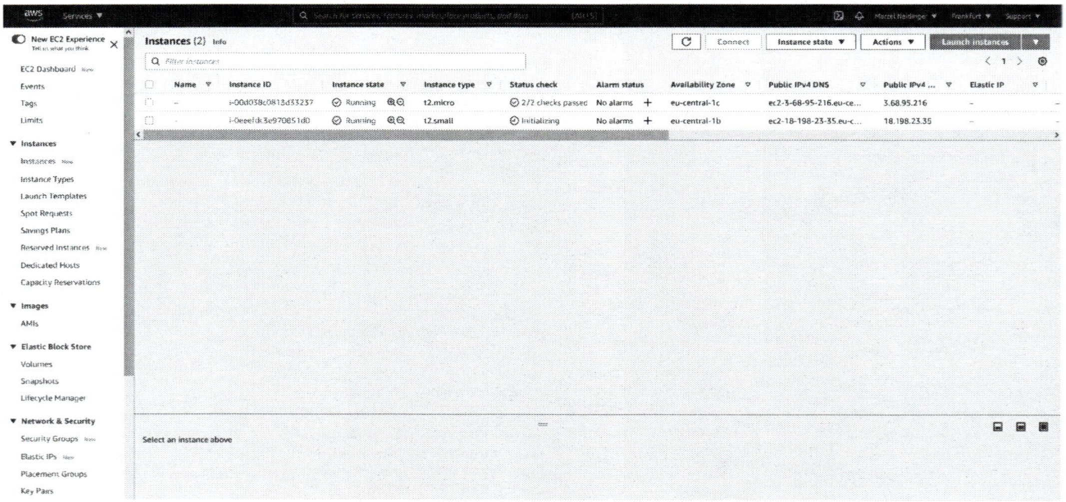

Figure 11.7 – Overview of running instances in the AWS console to verify that the new instance was created

The newly created instance should show up with a **Status check** of **Initializing**.

How it works...

So far, in this chapter, we have created a new EC2 instance programmatically. With our `boto3` client authenticated, we created a new EC2 instance. In the previous recipe, *Collecting information about your cloud networking resources*, we retrieved a client instead of a resource. The difference between the two is the level of abstraction. A client is a low-level abstraction that is shared between the SDK and the AWS CLI tool. A client maps all the operations provided by the AWS API one to one and returns the JSON information that's retrieved from the API. A resource, on the other hand, is a boto3-specific abstraction that builds on top of a client and exposes the information in an object-oriented format. However, not all operations and services are available as resources. If you want to access something that is not present as a resource, you'll have to fall back to using a client instead. With our EC2 resource retrieved, we used the `create_instances()` function to create one or more EC2 instances. To do so, we needed to provide an `ImageId`. This `ImageId` specifies which image to use as a basis for the newly created instance. Some of the most prominent image ID are as follows:

- Amazon Linux 2 AMI, a general-purpose Linux image with an ID of **ami-0bad4a5e987bdebde**.

- Red Hat Enterprise Linux 8 has an ID of **ami-0bad4a5e987bdebde**.
- SUSE Linux Enterprise Server 15 SP2 (HVM) has an ID of **ami-09e8a19c9eda495b3 (64-bit x86)**.
- Ubuntu Server 20.04 LTS has an ID of **ami-05f7491af5eef733a**.
- Microsoft Windows Server 2019 Base has an ID of **ami-086d0be14ab5129e1**.

You can find a full list of available image IDs by accessing the AWS console, navigating to the EC2 service, and then starting the process of creating a new EC2 instance:

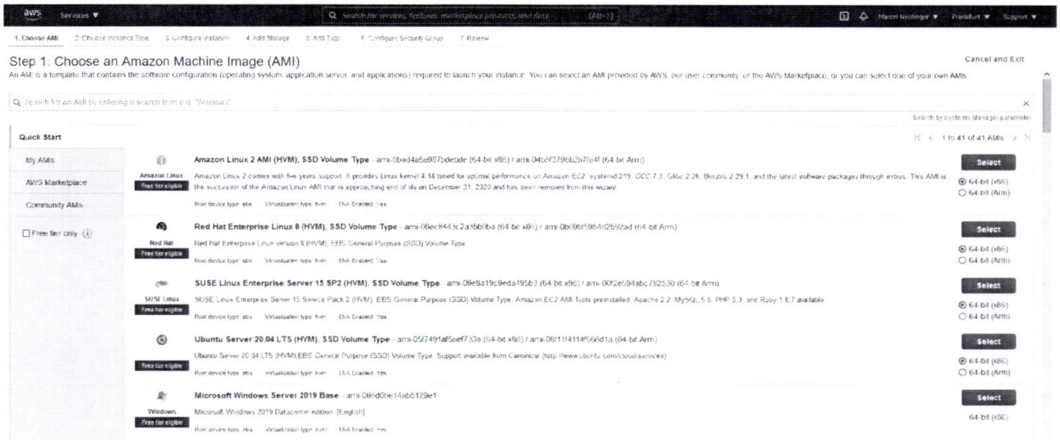

Figure 11.8 – Overview of the available images in the console

Besides the officially supported images, there is also the AWS marketplace, where images can be published. The filters on the left allow you to search for images that can be used with the Free Tier, which is restricted to some images.

Creating a VPC

VPC is an AWS service that lets you create separate and isolated environments for your resources to run in. In fact, every EC2 instance you start is part of your default VPC, but creating and then associating an EC2 instance with a specific VPC allows you to have full control over routing tables, internal IP addresses, and all other aspects of your instance's connectivity. In this recipe, we will learn how to create such a VPC programmatically. In the *Subnetting your VPC* recipe, we will create a subnet within a VPC and create an EC2 instance in the previously created VPC. Then, in the *Changing routes in your VPC* recipe, we'll learn how to change the routes in our VPC. But before we can change these properties, we'll have to create a VPC.

Getting ready

Open your code editor and start by creating a file called `create_vpc.py`. Next, navigate your terminal to the same directory that you just created the `_vpc.py` file in.

You'll have to have authentication set up as described in the *Setting up the library to interact with your AWS account* recipe.

How to do it...

Follow these steps to create a VPC programmatically:

1. Import the `boto3` library:

```
import boto3
```

2. Create a new EC2 resource. VPCs are logically grouped in the EC2 section:

```
ec2_resource = boto3.resource('ec2')
```

3. Create a new VPC for the `10.10.0.0/16` CIDR block:

```
vpc = ec2_resource.create_vpc(CidrBlock='10.10.0.1/16')
```

4. Wait until the VPC resource has been created:

```
vpc.wait_until_available()
```

5. Finally, print out the ID of our newly created VPC:

```
print(f"Created vpc {vpc.id}")
```

6. Run the script by typing the following command:

```
python3 create_vpc.py
```

The output of your script should look as follows. The ID of your VPC will be different:

Figure 11.9 – Output of our script for creating a VPC

You can verify that the VPC has been created properly by accessing the AWS console on the web:

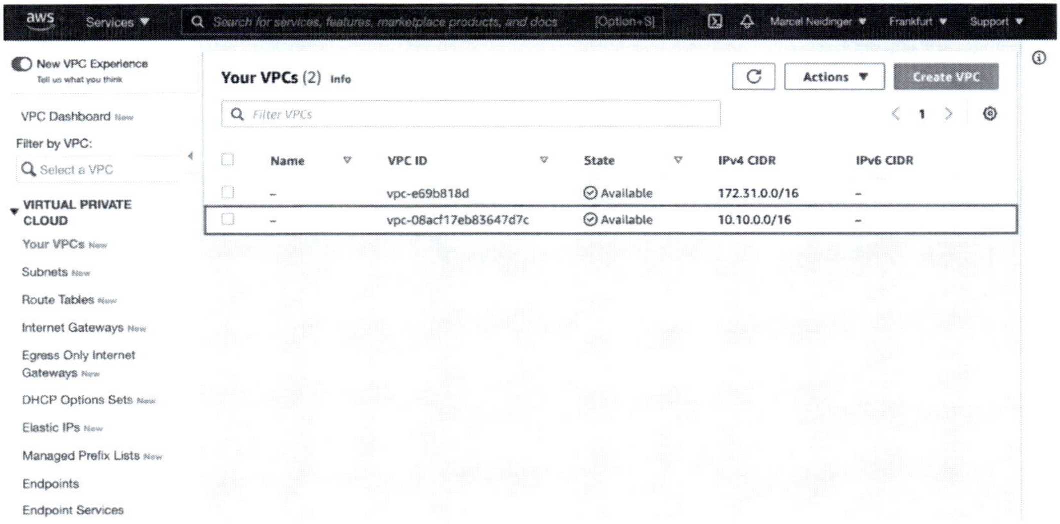

Figure 11.10 – The AWS console showing our newly created VPC

How it works...

With our successfully authenticated `boto3` library (see the *Setting up the library to interact with your AWS account* recipe for an explanation on how to do this), we created an EC2 resource. Compared to the client that we saw in previous recipes, a resource is a high-level abstraction that provides additional abstractions and functionalities. We used the resources for the `create_vpc()` function to retrieve a newly created VPC. This function returns a VPC Python object that provides the `wait_until_available()` helper function. This function will pause the execution of our script until the VPC resource has been created successfully and is available. Finally, we printed out the ID of our VPC, an alphanumeric string that always starts with `vpc-`, back to the user. With this, we navigated to the VPC section in our EC2 web console and validated that our VPC has indeed been created with the CIDR block we specified.

There's more...

There is a maximum number of VPCs that you can create in your account:

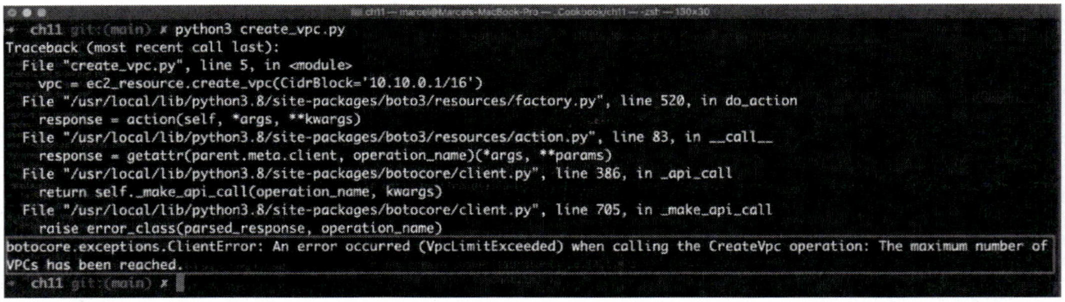

Figure 11.11 – Error message returned by boto3 when too many VPCs already exist

`boto3` will inform you of this by throwing a `ClientError`, which contains an error message indicating that too many VPCs exist. If you encounter this error, you'll have to delete one of your VPCs, either programmatically or via the AWS console.

Subnetting your VPC

VPCs already separate your resources in the cloud into logical and separated blocks but, within your VPC, you can have an additional separation in the form of subnets. Subnets allow you to further specify the routing and accessibility of your EC2 resources.

In this recipe, you are going to learn how to create a VPC and, in that newly created VPC, create a new subnet. Within this subnet, you'll then create an EC2 instance.

Getting ready

Open your code editor and start by creating a file called `subnet_vpc.py`. Next, navigate your terminal to the same directory that you just created the `subnet_vpc.py` file in.

You'll have to have the same authentication that you set up in the *Setting up the library to interact with your AWS account* recipe.

Additionally, you will need a previously set up SSH key pair. This is not the key pair that we used for authenticating to the API, but rather an SSH key that can be used to log into the instance itself via SSH once launched. You can create this key pair in the console. You can find a guide on how to do this at `https://docs.aws.amazon.com/AWSEC2/latest/UserGuide/ec2-key-pairs.html`.

How to do it...

Follow these steps to create a new subnet and start an EC2 instance within it:

1. Import the `boto3` library:

```
import boto3
```

2. Define the key pair and image ID that our EC2 instance will use:

```
KEY_PAIR = "<Insert your key pair name here>"
IMAGE_ID = "ami-05f7491af5eef733a"
```

3. Next, create an `ec2` resource with our `boto3` library:

```
ec2_resource = boto3.resource('ec2')
```

4. Then, create a new VPC that our subnet will be a part of:

```
vpc = ec2_resource.create_vpc(CidrBlock='10.10.0.0/16')
vpc.wait_until_available()

print(f"Created vpc {vpc.id}")
```

5. Next, create a new subnet. The subnet's CIDR block needs to be a part of the IP block we've defined for our VPC:

```
subnet = vpc.create_subnet(CidrBlock='10.10.1.0/24')
```

6. With our subnet created, we can define the network interface of our EC2 instance, which will be part of the newly created subnet:

```
net_inf = []
net_inf.append({
    'SubnetId': subnet.id,
    'DeviceIndex': 0
})
```

7. With the interface defined, we can go ahead and create our EC2 instance:

```
i = ec2_resource.create_instances(ImageId=IMAGE_ID,
                                  KeyName=KEY_PAIR,
                                  NetworkInterfaces=net_
inf,
```

```
                                    MaxCount=1,
                                    MinCount=1
                                    )
```

8. Finally, print out the ID of our instance:

```
print(f"Instance with id {i[0].id} created")
```

9. Run the script by typing the following command:

```
python3 create_vpc.py
```

The output of your script should look as follows. The ID of your created VPC and EC2 instance will be different:

Figure 11.12 – Output of our subnetting script

We can verify that our script worked by navigating to the `EC2` page of the AWS console. By clicking on the newly created instance, whose ID we have seen in the output of our script, we can access the instance summary:

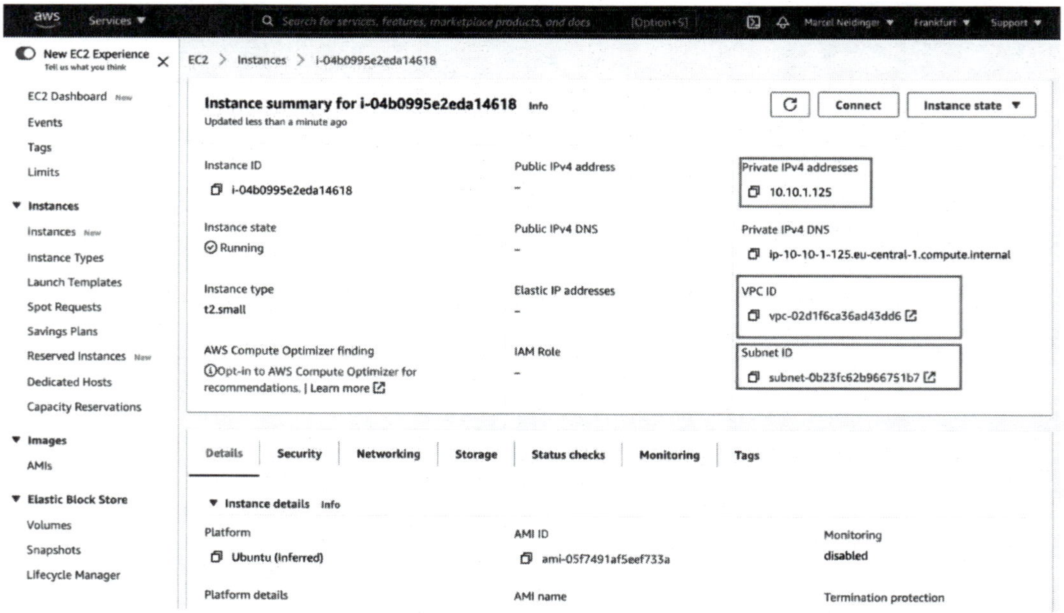

Figure 11.13 – Instance summary for our automatically created EC2 instance

As you can see, our instance has a private **IPv4 address** from the subnets block and has associations with the VPC that was created, as well as the subnet that we created.

How it works...

With our `boto3` library authenticated (see the *Setting up the library to interact with your AWS account* recipe for an explanation of how to do this), we defined the properties of the EC2 instance we wanted to create. Specifically, we needed to define the key pair and the image that will be used for this instance. You can find a more detailed explanation of these two properties in the *How it works...* section of the *Starting EC2 instances* recipe.

With these properties defined, we created a VPC with a defined CIDR block and, within that VPC, created a new subnet that is part of the VPC's CIDR block. We then defined a new interface for our EC2 instance. Since an EC2 instance can have multiple (virtual) network interfaces, we needed to provide a list here that we can append a dictionary that specifies each of our interfaces to. In this example, we only created a single virtual network interface with a device ID of `0`, and we referenced the subnet ID of the previously created subnet to associate this interface with the subnet. After creating our EC2 instance, this prompted AWS to assign an IP from that subnets pool to this interface. With the interface information defined, we used the `boto3` library to create the EC2 instance. This process is the same as what we saw in the *Starting EC2 instances* recipe, with the exception that we passed the `network interfaces` parameter as well, which prompts AWS to create the previously defined virtual networking interface for our instance.

With the instance created, we printed out the ID. This allows us to find it more easily in the AWS console, to verify that the EC2 instance has indeed been created in our VPC and has been assigned an IP address from the subnets pool.

You can find more information on how to implement subnets in AWS in the official documentation: `https://docs.aws.amazon.com/vpc/latest/userguide/VPC_Subnets.html`.

Changing routes in your VPC

With our VPCs and subnets created programmatically, we can now think about routing and, more specifically, how we want to route from a subnet to different destinations. AWS uses the concept of routing tables and routes, all of which can be configured programmatically. In this recipe, we are going to create a new VPC, associate a subnet with it, define an internet gateway for that subnet, and then change the routing table of our VPC and subnet so that it uses the previously created gateway.

Getting ready

Open your code editor and start by creating a file called `create_routes.py`. Next, navigate your terminal to the same directory that you just created the `create_routes.py` file in.

You'll have to have the same authentication that you set up in the *Setting up the library to interact with your AWS account* recipe.

How to do it...

Follow these steps to create a new route in your VPC:

1. Import the `boto3` library:

```
import boto3
```

2. Let's start by defining the destination CIDR that we want our route to be valid for:

```
DST_CIDR = "0.0.0.0/0"
```

3. Next, create an `ec2` resource from `boto3`:

```
ec2_resource = boto3.resource('ec2')
```

4. Create a new VPC and print out the VPC's ID for debugging:

```
vpc = ec2_resource.create_vpc(CidrBlock='10.10.0.0/16')
vpc.wait_until_available()
print(f"Created vpc {vpc.id}")
```

5. Create a new subnet and print out that ID as well:

```
subnet = vpc.create_subnet(CidrBlock='10.10.1.0/24')
print(f"Created subnet {subnet.id}")
```

6. Define a new gateway and attach the newly created internet gateway to our VPC:

```
gateway = ec2_resource.create_internet_gateway()
vpc.attach_internet_gateway(InternetGatewayId=gateway.id)
print(f"Created gateway {gateway.id} and associated with
{vpc.id}")
```

7. Create a new route table within our VPC and associate the route table with our previously created subnet:

```
route_table = ec2_resource.create_route_table(VpcId=vpc.
id)
route_table.associate_with_subnet(SubnetId=subnet.id)
print(f"Created route table {route_table.id}")
```

8. Finally, create a new route from our `Destination` block and associate the internet gateway we created previously:

```
r = route_table.create_route(DestinationCidrBlock=DST_
CIDR, GatewayId=gateway.id)
print(f"Created route {r} for {DST_CIDR} block")
```

9. Run the script by typing the following command:

```
python3 create_routes.py
```

The output of your script should look as follows. The IDs of your created VPC, subnet, gateway ID, and route table will be different:

Figure 11.14 – Output from our route creation script

You can verify that the resources have been created correctly by navigating to the VPC's Details page in the AWS console, as shown in the following screenshot:

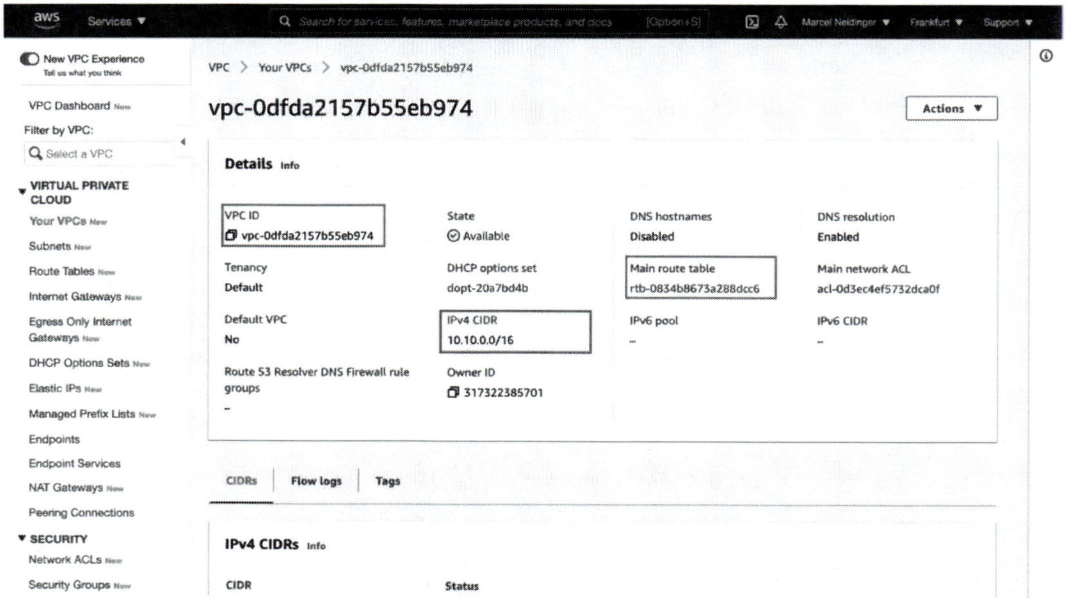

Figure 11.15 – VPC details showing the VPC's CIDR block and routing table that we created

How it works...

With our `boto3` library authenticated (see the *Setting up the library to interact with your AWS account* recipe for an explanation of how to do this), we defined the destination CIDR block for our new rules. Next, we retrieved an `ec2` resource from `boto3`. This resource is an abstraction that's built on top of the JSON data returned by the AWS API gateway itself, which allows us to deal with the API in a more abstracted way. As an example, the function to create a VPC, `create_vpc()`, does not return a pure dictionary but rather a Python object with additional helper functions such as the wait function, which halts the execution of our script until the VPC has been created. With the VPC for our CIDR block created (see the *Creating a VPC* recipe, for a more detailed explanation) we created a new subnet within that VPC.

For our newly created subnet, we created a new routing table that we assigned to the subnet by using the `associate_with_subnet()` function. With our route table created and associated, we created a new route from the destination CIDR block we defined at the beginning of the script to the internet gateway we just created. Finally, after running the script, we verified that it worked by accessing the VPC's details in the EC2 section of the AWS console.

12
Automating your Network Security Using Python and the Firepower APIs

Security has become a central concern for every network operations team in the world. Whether you are securing the network of a global corporation or the wireless access of your local coffee shop, network security touches every aspect of network operations.

In this chapter, we are going to focus on the automation abilities of Cisco's **Firepower Management Center** (**FMC**) by using its **Representational State Transfer** (**REST**) **application programming interface** (**API**). Using automation together with FMC allows us to constantly check our enacted access policies or automatically update them.

In this chapter, we are going to cover the following recipes:

- Exploring the API Explorer
- Authenticating against the FMC REST API
- Retrieving access policies
- Changing access policies
- Retrieving access rules
- Changing access rules
- Deleting access rules

Technical requirements

For this section and the remainder of the book, you'll need an installation of Python. Specifically, you'll need a Python interpreter of version 3.6.1 or higher. This book makes use of language constructs of Python 3 and thus is incompatible with Python 2.x. Additionally, you'll need to install the `requests` package. You can install the newest version of `requests` by using `python3 -m pip install requests`. At the time of this writing, the current version is version 2.25.1.

You will also need a code editor. Popular choices include Microsoft **Visual Studio Code** (**VS Code**) or Notepad++.

Additionally, you'll need access to a Cisco FMC instance that has the REST API enabled, as well as a valid user account. Please have a look at the reference guide for your FMC installation on how to enable the REST API. You can find the reference guide at this link: `https://www.cisco.com/c/en/us/support/security/defense-center/products-programming-reference-guides-list.html`.

The scripts in this recipe have been tested against version 6.3 of FMC.

You can view this chapter's code in action here: `https://bit.ly/3CL6z1c`

Exploring the API Explorer

When diving into a new API, it can be quite daunting to explore it. While trying out requests from a Python script can be done, it is usually easier to explore in a **graphical user interface** (**GUI**). Many cloud APIs offer this type of interactive documentation where, with a few clicks, you can not only see the schemas of a request, which **Uniform Resource Locators** (**URLs**) to use, and which parameters to pass but can also run the request directly from your web browser. FMC offers a similar feature on the device itself, called the **API Explorer**. Based on the *OpenAPI Specification*, a standard to define REST APIs, an interactive page for the API can be generated and used to execute API requests from within your browser.

In this recipe, we are going to have a look at the FMC API Explorer and explore its capabilities:

1. Access your FMC installation's API Explorer by navigating to the following link: https://<path_to_your_fmc_webinterface> /api/api-explorer/.

   After authenticating with your username/password combination, you will be presented with an overview of all the available API endpoints:

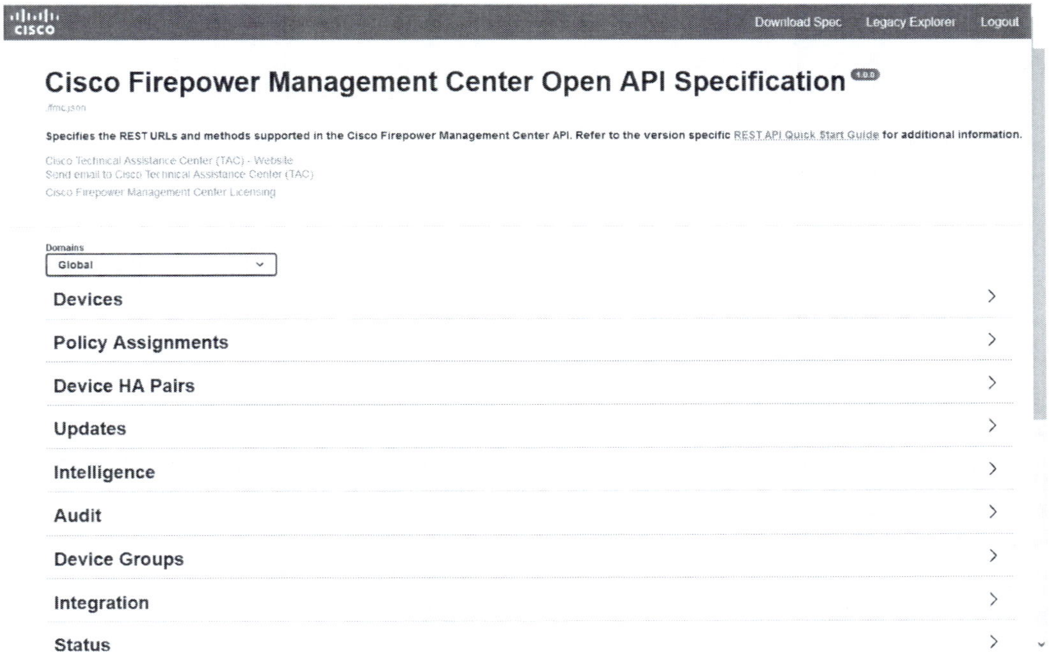

Figure 12.1 – Overview of the available FMC API endpoints

2. Click on one of the endpoints—for example, **System Information**. The expanded section contains all requests that can be made within that endpoint. The URL, as well as the required request method—in this case, `GET`—are specified:

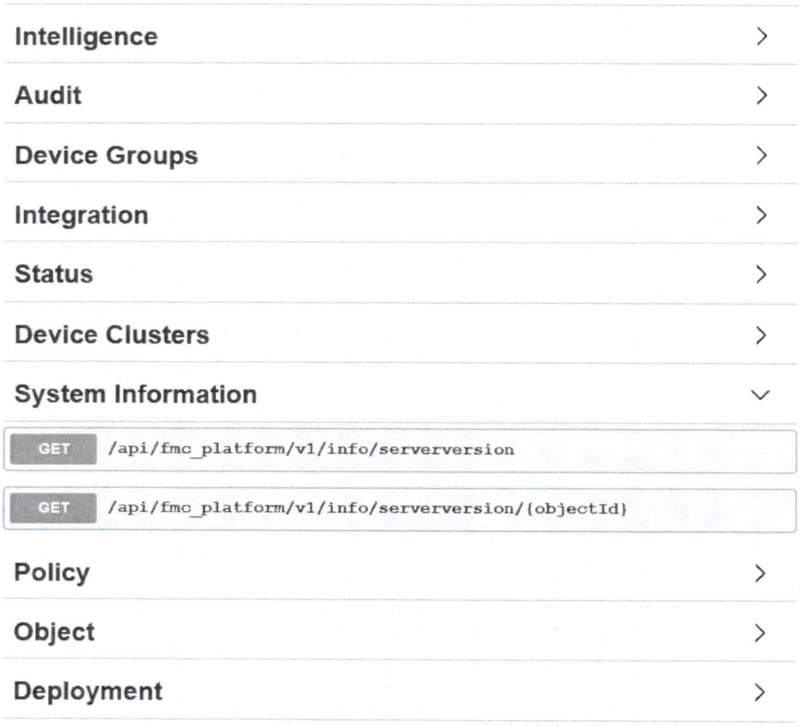

Figure 12.2 – List of all available URLs under the System Information endpoint

3. Click on one of the endpoints— for example, the one with the `/api/fmc_platform/v1/info/serverversion` URL. On the resulting page, you have information on the parameters that can be passed to the endpoint. In this case, these are the following:

   a) **offset**, which is of type **integer** and is part of the **query** parameters.

   b) **limit**, which is of type **integer** and is part of the **query** parameters.

   c) **expanded**, which is of type **Boolean** and is part of the **query** parameters.

   d) Additionally, each of the parameters has a description.

Below the parameters, there is a section for all the possible responses grouped by their response code, as well as an exemplar response body. You can use this example to understand what kind of information is returned by the API and how this is formatted:

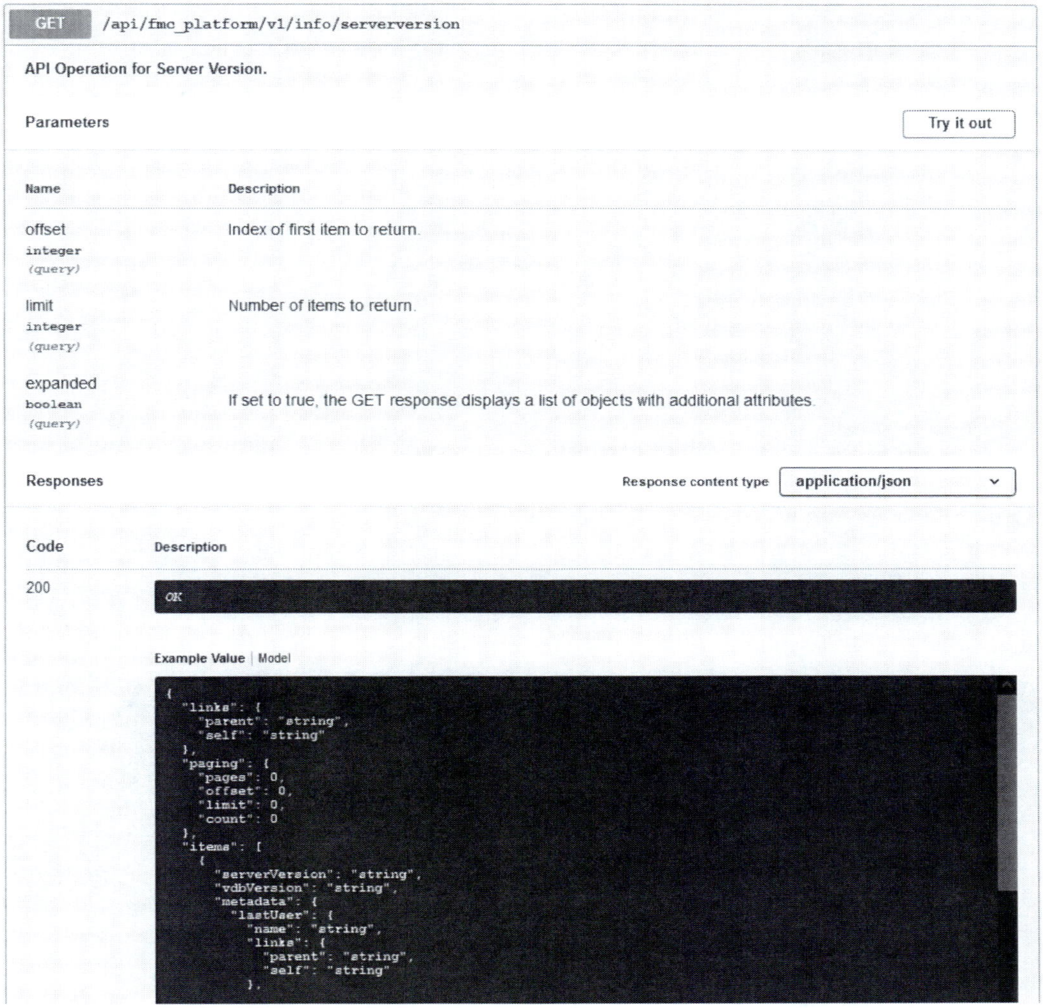

Figure 12.3 – Overview of a single request in the API

4. Click on the **Try it out** button in the top-right corner. This will open a new dialog to try out the request. The input fields for the query parameters allow you to specify values for each of these parameters. We can then execute the request by clicking on the blue **Execute** button:

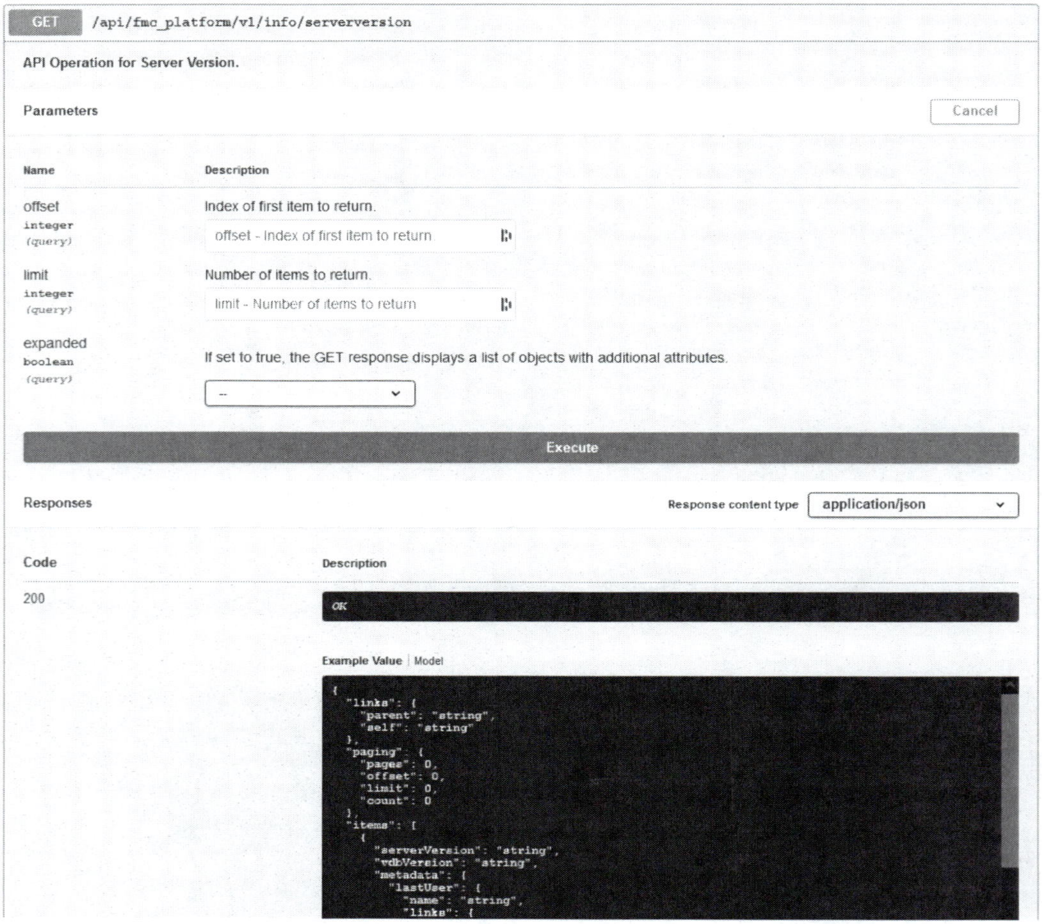

Figure 12.4 – The Try it out page for your chosen API endpoint

5. After clicking on **Execute**, you will be presented with the result of your request. There are two fields: one for the response body—in this case, **JavaScript Object Notation** (**JSON**)-formatted information about the version of our FMC server—and another for the response headers that contain all the meta-information returned in the header of our request:

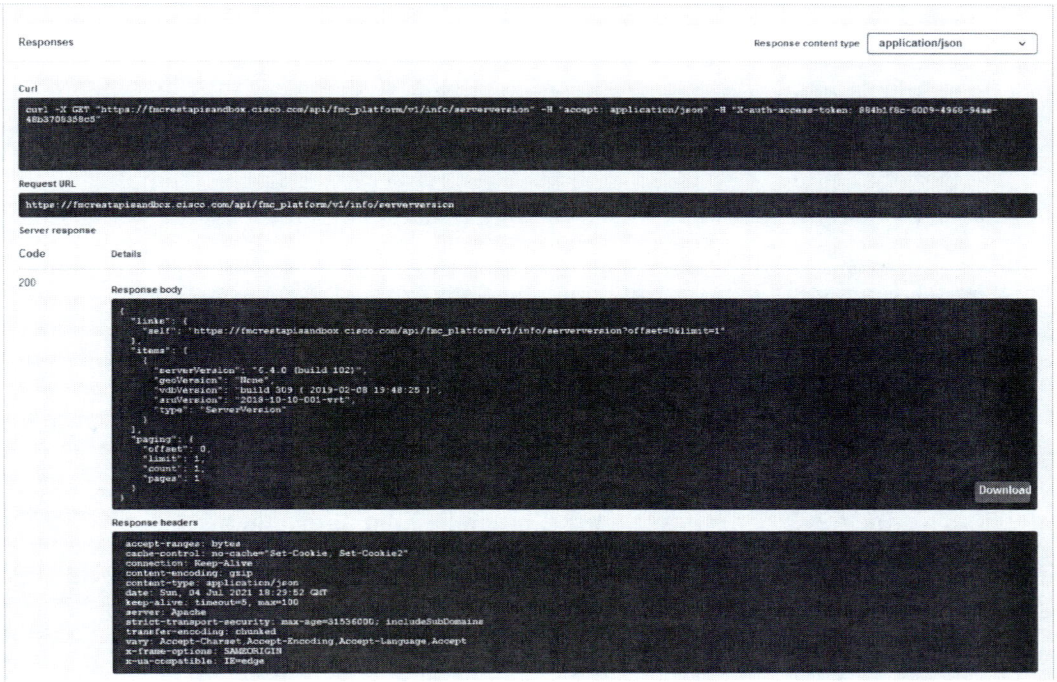

Figure 12.5 – Response after executing our query from the API Explorer

Using the API Explorer, and interactive API documentation in general, it's easier to try out a new API from within your browser. If you want to check whether other APIs of products you are using have interactive API documentation, look out for names such as the **OpenAPI Specification**, **Swagger Docs** (Swagger is an old name for the format and tools used to describe a REST API and generate these interactive docs), or **interactive documentation**.

Authenticating against the FMC REST API

In this recipe, we'll learn how to obtain an authentication token for our FMC console. In contrast to other REST APIs, such as the Cisco **Meraki API** we explored in a previous chapter, FMC's REST API does not use a static authentication token that can be generated for a user and is valid indefinitely. Instead, we must use our username/password combination to obtain a token that can then be used for all further requests to authenticate the request. In addition to the token itself, the response also contains a list of available domains that our account has access to.

Since we must do this authentication for every request, we'll write a function that takes our username/password combination as well as the base URL of our console and returns an authenticated `requests session` object, as well as a list of domains. We will then reuse this function in the coming chapters to create an authenticated session.

Getting ready

Open your code editor and start by creating a file called `authenticate.py`. Next, navigate in your terminal to the same directory in which you just created the `authenticate.py` file.

How to do it...

Follow these steps to authenticate against the FMC REST API:

1. Import the required `requests` library as well as the `json` and `sys` modules from the standard library:

```
import requests
import json
import sys
from requests.auth import HTTPBasicAuth
```

2. Next, specify the base URL as well as our login information:

```
base_url = "https://fmcrestapisandbox.cisco.com"

username = "<Insert username here>"
password = "<Insert password>"
```

3. Next, we create our authentication function:

```
def get_authenticated_session(user, password, base_url):
```

4. Within this authentication function, specify the authentication URL and authentication details:

```
    auth_url = f"{base_url}/api/fmc_platform/v1/auth/
generatetoken"
    auth = HTTPBasicAuth(user, password)
```

5. Send a POST request with an empty body to the auth URL. Depending on your FMC installation, you'll have a self-signed certificate. If that is the case, set the verify property of the POST request to False. This will prevent requests from verifying the **Secure Sockets Layer** (**SSL**) certificate:

```
    resp = requests.post(auth_url, auth=auth,
verify=False)
```

6. Next, we create a new session object and, if the request to the authentication URL was successful, set the authentication header of our session based on the token returned from the request to the authentication URL:

```
    s = requests.Session()
    if resp.ok:
        s.headers.update({
            'X-auth-access-token': resp.headers.get('X-
auth-access-token')
        })
```

7. Next, we disable **SSL verification** for our session, print out a message that our authentication was successful, and parse the domains. These are **JSON-encoded** into the DOMAINS header field:

```
        s.verify = False
        print("Authenticated successfully!")
        domains = json.loads(resp.headers.get('DOMAINS'))
```

8. To finish off, we return the session as well as our list of domains:

```
        return s, domains
```

9. If our request failed, we print out the status code returned from FMC and stop the execution of the script:

```
    else:
        print(f"Failed to authenticate. Response code:
{resp.status_code}")
        sys.exit(-1)
```

10. With our authentication function finished, we can use it to retrieve an authenticated session as well as a list of domains:

```
sess, domains = get_authenticated_session(username,
password, base_url)
```

11. And then, for each of our domains, we can print out the name and the **identifier** (**ID**):

```
for d in domains:
    print(f"{d['name']}: {d['uuid']}")
```

12. To run this script, go to your terminal and execute the following command:

```
python3 authenticate.py
```

The output of your script should look like this:

Figure 12.6 – Output from our authentication script

The name and ID of your domain(s) will be different.

How it works...

In this recipe, we first import the required libraries. We will use the `requests` module again to make our **HyperText Transfer Protocol** (**HTTP**) requests, and we also use the `sys` and `json` modules from the standard library. After specifying our credentials (see the *There's more...* section of this recipe for hints on how to do this in production code), we start by defining our authentication function. This function takes three arguments: `username`, `password`, and the base URL of our FMC console. Within the function, we use our base URL to construct the authentication URL and create HTTP **basic authentication** (**HTTP Basic Auth**) details using our username and password. With this pre-work done, we send an empty `POST` request to our authentication URL that contains the username/password combination as an `HTTP Basic Auth` header. For checking whether our request was successful or not, we use the `ok` property of our response object. This Boolean property will be `True` if the status code of our response was below `400`. If this is not the case, we print out the response code of our failed request and end the execution of our script using the standard library's `sys` module and the `exit` function.

If the request succeeded, we extract the authorization token from the response header. FMC uses an `X-auth-access-token` header key instead of the more common `Authorization` key. We copy this key into the headers of our session object and disable the SSL verification for our session. The last step is only required if your FMC installation uses a self-signed certificate. In addition to the token, our response headers also contain a JSON-encoded list of domains in the `DOMAIN` response-header field. We can decode the associated `JSON` and retrieve a list of domains. To finish our function, we then return both the session and domain.

Next, we use our previously written function to retrieve the authenticated session and domains and iterate over all of our domains to print out the name and ID.

There's more...

In production code, you should not store your credentials in a script. You can use the following snippet to read the username and password from your environment variables:

```
import os

username = os.environ.get('FMC_USERNAME')
password = os.environ.get('FMC_PASSWORD')
```

Before running your scripts, you will then have to first set the environment variables. In Linux/macOS X, you can do this by using the `export` command:

```
export FMC_USERNAME="<insert username here>"
export FMC_PASSWORD="<insert password here>"
```

If you are using Linux or Mac OS X, you can also make this permanent by adding it to your `.bashrc` or `.profile` file.

Retrieving access policies

Access policies configured in your FMC instance can easily be retrieved using the FMC REST API. However, doing `GET` requests to list large amounts of data can be computationally expensive for the API server. Imagine you have an API that returns you a list of all the policies in your network. This list might have, as an example, 5,000 entries, and retrieving all the information is computationally expensive. Let's imagine the API would always return a full list of entries. If your script now searches for one specific policy, based on the name, and that entry was the first entry in your list, you would have requested the information about the other 4,999 access policies without using the information. To lessen the load on API servers, most REST APIs employ a concept called **pagination** when dealing with large lists of items.

Instead of always returning all entries, they return you a selection—let's say 50 entries. This response is called a **page**. In addition to the entries themselves, the page also contains information about how to request the next page. This means that, for requesting all our 5,000 network policies from the preceding example, we do not send one large request but instead, we send a first request that gives us the first 50 entries, as well as information about where to get the next page, which contains the next 50 entries. Using this principle, if we want all 5,000 entries, we do 100 small requests with 50 items each, instead of one large request that contains all the items. Pagination is a very common and useful concept employed by most APIs for load balancing, and the FMC REST API uses pagination as well. In this recipe, you'll see how we can retrieve a list of all access policies that are configured in a domain and resolve the resulting pagination.

Getting ready

Open your code editor and start by creating a file called `get_access_policies.py`. Next, navigate in your terminal to the same directory in which you just created the `get_access_policies.py` file.

How to do it...

Follow these steps to retrieve your access policies from the FMC REST API:

1. Import the required `requests` library as well as the `json` and `sys` modules from the standard library:

```
import requests
import json
import sys
from requests.auth import HTTPBasicAuth
```

2. Specify the required variables such as your username, password, and domain ID:

```
base_url = "https://fmcrestapisandbox.cisco.com"

username = "<Insert username here>"
password = "<Insert password here>"
domain = "<Insert domain id here>"
```

3. Define an authentication function and use this function to obtain an authenticated session based on your username and password combination. Please have a look at the *How to do it…* and *How it works…* sections of the *Authenticating against the FMC REST API* recipe:

```
def get_authenticated_session(user, password, base_url):
    auth_url = f"{base_url}/api/fmc_platform/v1/auth/
generatetoken"
    auth = HTTPBasicAuth(user, password)

    resp = requests.post(auth_url, auth=auth,
verify=False)

    s = requests.Session()
    if resp.ok:
        s.headers.update({
            'X-auth-access-token': resp.headers.get('X-
auth-access-token')
        })
        s.verify = False
        print("Authenticated succesfully!")
        domains = json.loads(resp.headers.get('DOMAINS'))

        return s, domains
    else:
        print(f"Failed to authenticate. Response code:
{resp.status_code}")
        sys.exit(-1)

sess, domains = get_authenticated_session(username,
password, base_url)
```

4. Define a URL that we want to request our policies from, as well as an empty list of items. Then, do the request itself using our previously obtained session:

```
url = f"{base_url}/api/fmc_config/v1/domain/
{domain}/policy/accesspolicies"

items = []
resp = sess.get(url)
```

5. If the request was successful, we first decode the returned data and add all the items—in this case, dictionaries representing access policies—to the `items` list:

```
if resp.ok:
    data = resp.json()

    for i in data['items']:
        items.append(i)
```

6. Next, we need to take care of all the additional pages. We define an empty `next_links` list and then extract the links to the remaining pages from the first response:

```
next_links = []

    if "next" in data['paging']:
        if isinstance(data['paging']['next'], list):
            next_links = data['paging']['next']
        else:
            next_links.append(data['paging']['next'])
```

7. With the list of remaining pages retrieved, we can then iterate over this list and send a `GET` request for each of these links:

```
    for link in next_links:
        print(f"Requesting url '{link}'")
        r = sess.get(link)
```

8. If this request was successful, we add all the access policies contained within the `items` list. If the request was unsuccessful, we print out the status code, as well as the link that our request failed on:

```
        if r.ok:
            for i in r.json()['items']:
                items.append(i)
        else:
            print(f"Failed to request url '{link}'.
Status code: {r.status_code}")
```

9. Finally, we need to handle the case that our initial request failed and then print out the name and ID of all of our access policies by iterating over the `items` list:

```
else:
    print(f"Failed to request url '{url}'. Status code:
{resp.status_code}")

for i in items:
    print(f"{i['name']}: {i['id']}")
```

10. To run this script, go to your terminal and execute the following command:

```
python3 get_access_policies.py
```

The output of your script should look like this:

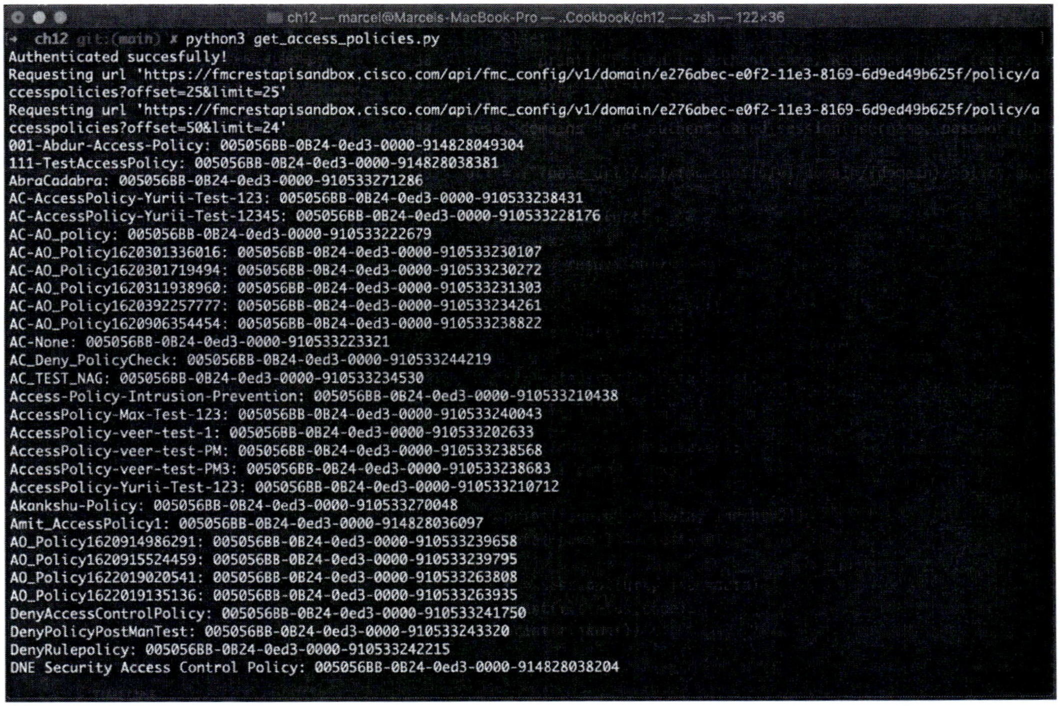

Figure 12.7 – Output from our access policy retrieval script

The name and ID of your policies will be different. Note how, at the top, there are two additional pages that we are requesting, denoted by the `Requesting url` debug message.

How it works...

We start by importing our required libraries, and then define our connection details such as the username/password combination to use for authentication, the base URL of our FMC instance, and the domain under which we are operating. Next, we define our authentication function. For a more detailed explanation of how the authentication process works, please refer to the *How to do it...* and *How it works...* sections of the *Authenticating against the FMC REST API* recipe.

With our authenticated session, we can do our initial request. Remember that for a paginated result, the response of this request will contain at least two pieces of information:

- A container—in the case of FMC's REST API, this is called `items`. Contained within this list are the items—in our case, dictionaries representing access policies—that we are requesting.
- A pagination section—in the case of FMC's REST API, this is called `paging`— that contains information about the next and previous pages.

We first copy all items from the `items` key in our response into a local list called `items`. We then retrieve a list of the next URLs. This list contains URLs to all the other pages that are part of this response. We iterate over this list and do another `GET` request for each of them. We then parse the result from our request and add it to our local `items` list. Once we have requested all pages and stored the results in our local `items` list, we can then go ahead and iterate over them, printing out the name and ID.

Changing access policies

Now that we have seen how to retrieve information about our access policy, what about changing it? Maybe we want to automatically carry out a workflow that checks all access policies against a list of valid policies and either automatically deactivates the policy or at least marks it for someone to have a look at it. We have seen the automatic retrieval of all lists in the previous recipe, *Retrieving access policies*. This recipe will cover the part about changing an access policy based on the policy ID.

Getting ready

Open your code editor and start by creating a file called `change_access_policies.py`. Next, navigate in your terminal to the same directory in which you just created the `change_access_policies.py` file.

How to do it...

Follow these steps to change one of your access policies using the FMC REST API:

1. Import the required `requests` library as well as the `json` and `sys` modules from the standard library:

```
import requests
import json
import sys
from requests.auth import HTTPBasicAuth
```

2. Specify the required variables such as your username, password, domain ID, and the ID of the policy you want to change:

```
base_url = "https://fmcrestapisandbox.cisco.com"

username = "<Insert username here>"
password = "<Insert password here>"
domain = "<Insert domain id here>"
policy_id = "<Insert your policy id here>"
```

3. Define an authentication function and use this function to obtain an authenticated session based on your username and password combination. Please have a look at the *How to do it...* and *How it works...* sections of the *Authenticating against the FMC REST API* recipe:

```
def get_authenticated_session(user, password, base_url):
    auth_url = f"{base_url}/api/fmc_platform/v1/auth/
generatetoken"
    auth = HTTPBasicAuth(user, password)

    resp = requests.post(auth_url, auth=auth,
verify=False)

    s = requests.Session()
    if resp.ok:
        s.headers.update({
            'X-auth-access-token': resp.headers.get('X-
auth-access-token')
        })
```

```
        s.verify = False
        print("Authenticated succesfully!")
        domains = json.loads(resp.headers.get('DOMAINS'))

        return s, domains
    else:
        print(f"Failed to authenticate. Response code:
{resp.status_code}")
        sys.exit(-1)

sess, domains = get_authenticated_session(username,
password, base_url)
```

4. Specify the URL of the policy we want to change and request the initial data:

```
url = f"{base_url}/api/fmc_config/v1/domain/{domain}/
policy/accesspolicies/{policy_id}"

resp = sess.get(url)
```

5. If the request was successful, meaning that we have retrieved the details of our access policy, we can retrieve the data as a Python dictionary:

```
if resp.ok:
    data = resp.json()
```

6. Next, delete some of the API-specific keys in your data. This step is required so that we can update the resource using the information obtained from the `GET` request without running into API errors.
7. With the list of remaining pages retrieved, we can then iterate over this list and send a `GET` request for each of these links:

```
    if 'urls' in data.keys():
        del data['urls']

    if 'metadata' in data.keys():
        del data['metadata']

    if 'links' in data.keys():
        del data['links']
```

8. We can then change the property we want to change—in this case, our policy name—and post the data back to the API:

```
data['name'] = 'Test-API'

r = sess.put(url, json=data)
print(r.status_code)
print(r.json())
```

9. To run this script, go to your terminal and execute the following command:

```
python3 change_access_policies.py
```

The output of your script should look like this:

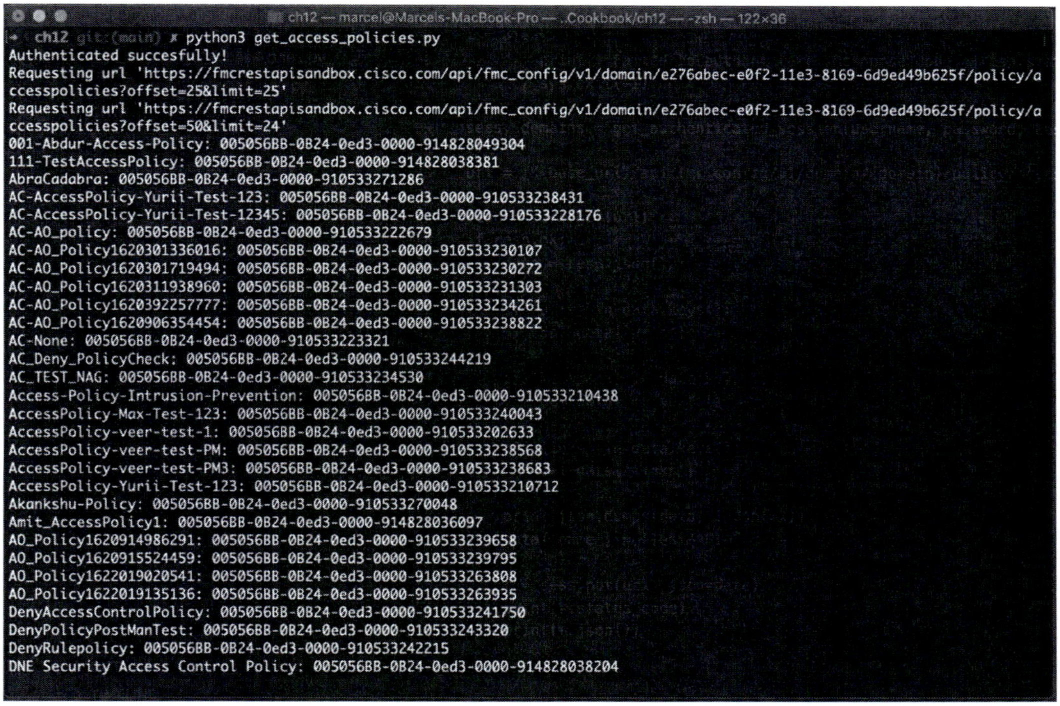

Figure 12.8 – Output from our access policy retrieval script

The data returned from the script will be different for your access policy.

How it works...

We start by importing our required libraries, and then define our connection details such as the username/password combination to use for authentication, the base URL of our FMC instance, and the domain under which we are operating. Next, we define our authentication function. For a more detailed explanation of how the authentication process works, please refer to the *How to do it...* and *How it works...* sections of the *Authenticating against the FMC REST API* recipe.

With our authenticated session, we can retrieve the details of our access policy. When doing `PUT` requests, we need to provide all the fields of the original item. `PUT` essentially replaces a resource on the server, and thus we need to give it all the information we have. We obtain this information by using a `GET` request on the URL specifying our access policy, as identified by the policy ID. We then delete some API-specific information such as the links and URLs. With the data manipulated, we can then apply the changes we want to make. In this case, we change the name of our policy and, together with the previously retrieved data, then send it back to the API using a `PUT` request.

Retrieving access rules

The access policy resource covered so far in this chapter has, as is very common with REST APIs, more attributes that refer (or *link*) to another object. Our access policy, for example, also has a list of associated access rules that define which zone is allowed or disallowed from sending traffic to another zone. In this recipe, we will cover the process of retrieving all associated rules for a given access policy.

Getting ready

Open your code editor and start by creating a file called `get_access_rules.py`. Next, navigate in your terminal to the same directory in which you just created the `get_access_rules.py` file.

How to do it...

Follow these steps to retrieve all access rules associated with your access policy:

1. Import the required `requests` library as well as the `json` and `sys` modules from the standard library:

```
import requests
import json
import sys
from requests.auth import HTTPBasicAuth
```

2. Specify the required variables such as your username, password, domain ID, and the ID of the policy we want to retrieve our access rules for:

```
base_url = "https://fmcrestapisandbox.cisco.com"

username = "<Insert username here>"
password = "<Insert password here>"
domain = "<Insert domain id here>"
policy_id = "<Insert your policy id here>"
```

3. Define an authentication function and use this function to obtain an authenticated session based on your username and password combination. Please have a look at the *How to do it…* and *How it works…* sections of the *Authenticating against the FMC REST API* recipe:

```
def get_authenticated_session(user, password, base_url):
    auth_url = f"{base_url}/api/fmc_platform/v1/auth/
generatetoken"
    auth = HTTPBasicAuth(user, password)

    resp = requests.post(auth_url, auth=auth,
verify=False)

    s = requests.Session()
    if resp.ok:
        s.headers.update({
            'X-auth-access-token': resp.headers.get('X-
auth-access-token')
        })
        s.verify = False
        print("Authenticated succesfully!")
        domains = json.loads(resp.headers.get('DOMAINS'))

        return s, domains
    else:
        print(f"Failed to authenticate. Response code:
{resp.status_code}")
        sys.exit(-1)
sess, domains = get_authenticated_session(username,
password, base_url)
```

4. Next, define the URL we want to request our access rules from and send off the request:

```
url = f"{base_url}/api/fmc_config/v1/domain/{domain}/
policy/accesspolicies/{policy_id}/accessrules"
resp = sess.get(url)
```

5. Since we are doing a GET request to list items again, our result will be paginated. We'll request all entries from the API and store them locally in a list called `items`. Please refer to the *How to do it...* section of the *Retrieving access policies* recipe for a detailed explanation of how the retrieval of paginated results works:

```
items = []
resp = sess.get(url)
if resp.ok:
          data = resp.json()

    for i in data['items']:
        items.append(i)
    next_links = []

    if "next" in data['paging']:
        if isinstance(data['paging']['next'], list):
            next_links = data['paging']['next']
        else:
            next_links.append(data['paging']['next'])

    for link in next_links:
        print(f"Requesting url '{link}'")
        r = sess.get(link)

        if r.ok:
            for i in resp.json()['items']:
                items.append(i)
        else:
            print(f"Failed to request url '{link}'.
Status code: {r.status_code}")
else:
    print(f"Failed to request url '{url}'. Status code:
{resp.status_code}")
```

6. With all our access rules retrieved, we can then iterate over them all. We then use the `self`-link that is included with each of the returned resources to request the details for each of our access rules and then print out these details, such as name, action, and whether the rule is enabled or not:

```
for i in items:
    url = i['links']['self']

    r = sess.get(url)
    if r.ok:
        rule = r.json()

        print(f"{rule['name']}({rule['id']}): Action:
{rule['action']} Enabled: {rule['enabled']}")
```

7. To run this script, go to your terminal and execute the following command:

```
python3 get_access_rules.py
```

The result should look like this:

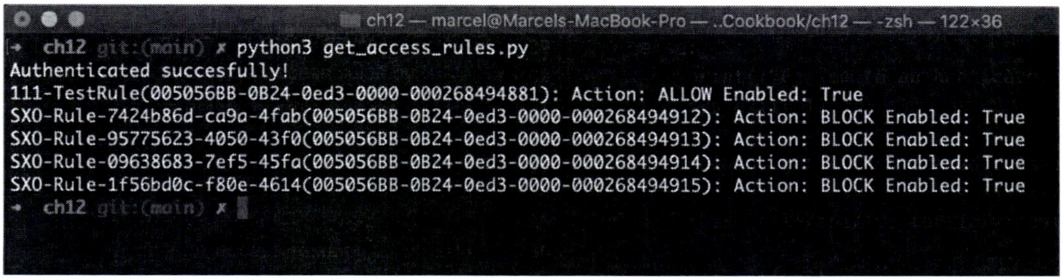

Figure 12.9 – Result of running our access rules retrieval script

The name of your access rules and their ID, action, and status may differ.

How it works...

We start by importing our required libraries, and then define our connection details such as the username/password combination to use for authentication, the base URL of our FMC instance, the domain under which we are operating, and the policy for which we want to retrieve the access rules. Next, we define our authentication function. For a more detailed explanation of how the authentication process works, please refer to the *How to do it...* and *How it works...* sections of the *Authenticating against the FMC REST API* recipe.

We then request all the configured access rules. These rules come as paginated results, and we thus have to resolve all the pages. Please refer to the *How it works...* section of the *Retrieving access policies* recipe for a more detailed explanation of how paginated results are retrieved.

With all our access rules retrieved, we can use another property that is specified by the API: the links section. Each object—in our case, this is a representation of an access rule—comes with a `links` property that contains the URLs. Specifically, under the name `self`, it contains a link to retrieve the details of this object. We could piece together this URL ourselves using the base URL, API prefix, domain ID, endpoint path, and access policy as well as the access rule ID, but using the `self`-link is a much more convenient way. After successfully requesting the details of each of our access rules, we print out the information obtained, such as the name, ID, action, and status.

Changing access rules

Now that we can automatically retrieve first the access policy and then its access rules, you might be wondering how we can change the rules themselves. A common task when dealing with access rules is to enable or disable them. In this recipe, we will see how you can update the properties of an existing access rule. In our case, we will disable the policy itself.

Getting ready

Open your code editor and start by creating a file called `change_access_rules.py`. Next, navigate in your terminal to the same directory in which you just created the `change_access_rules.py` file.

How to do it...

Follow these steps to change an existing access rule and disable it:

1. Import the required `requests` library as well as the `json` and `sys` modules from the standard library:

```
import requests
import json
import sys
from requests.auth import HTTPBasicAuth
```

2. Specify the required variables such as your username, password, domain ID, and the ID of the policy, as well as the ID of the rule we want to change:

```
base_url = "https://fmcrestapisandbox.cisco.com"

username = "<Insert username here>"
password = "<Insert password here>"
domain = "<Insert domain id here>"
policy_id = "<Insert your policy id here>"
rule_id = "<Insert your rule id here>"
```

3. Define an authentication function and use this function to obtain an authenticated session based on your username and password combination. Please have a look at the *How to do it...* and *How it works...* sections of the *Authenticating against the FMC REST API* recipe:

```
def get_authenticated_session(user, password, base_url):
    auth_url = f"{base_url}/api/fmc_platform/v1/auth/
generatetoken"
    auth = HTTPBasicAuth(user, password)

    resp = requests.post(auth_url, auth=auth,
verify=False)

    s = requests.Session()
    if resp.ok:
        s.headers.update({
            'X-auth-access-token': resp.headers.get('X-
auth-access-token')
        })
        s.verify = False
        print("Authenticated succesfully!")
        domains = json.loads(resp.headers.get('DOMAINS'))

        return s, domains
    else:
        print(f"Failed to authenticate. Response code:
{resp.status_code}")
```

```
        sys.exit(-1)
sess, domains = get_authenticated_session(username, 
password, base_url)
```

4. We start by retrieving the details of our access rule using a `GET` request to the URL that specifies our access rule resource based on the rule ID:

```
url = f"{base_url}/api/fmc_config/v1/domain/{domain}/
policy/accesspolicies/{policy_id}/accessrules/{rule_id}"

resp = sess.get(url)
```

5. If this request is successful, we can use the information to update our rule using a `PUT` request. We first retrieve the returned data and delete some API-specific properties such as the `links`, `metadata`, and `urls` sections:

```
if resp.ok:
    data = resp.json()
    if 'urls' in data.keys():
        del data['urls']

    if 'metadata' in data.keys():
        del data['metadata']

    if 'links' in data.keys():
        del data['links']
```

6. Now, we can change the property we want to change. In our example, this is going to be the `enabled` property, which we are going to switch to `False`:

```
    data['enabled'] = True
```

7. We can then send off a `PUT` request and check whether the request was successful or not. If it was not successful, we print out the response code; otherwise, we retrieve the returned access rule object itself and print out some information stating that the rule has been updated:

```
    r = sess.put(url, json=data)

    if r.ok:
        obj = r.json()
```

```
            print(f"Updated access rule {obj['name']}")
        else:
            print(f"Failed to update access rule. Status code: {r.status_code}")
```

8. Finally, we need to handle the case that our initial `GET` request failed:

```
else:
    print(f"Failed to retrieve access rule. Status code: {resp.status_code}")
```

9. To run this script, go to your terminal and execute the following command:

```
python3 change_access_rules.py
```

> **Important note**
>
> The name of your rule might differ.

The output of your script should look like this:

Figure 12.10 – Result of running our script to change an access rule

You can also verify that the change happened by running the script from the *Retrieving access rules* recipe again.

How it works...

We start by importing our required libraries, and then define our connection details such as the username/password combination to use for authentication, the base URL of our FMC instance, the domain under which we are operating, the policy for which our access rule is defined, and the ID of the rule itself. Next, we define our authentication function. For a more detailed explanation of how the authentication process works, please refer to the *How to do it...* and *How it works...* sections of the *Authenticating against the FMC REST API* recipe.

With our authenticated session, we first retrieve the details of the access rule. We need these details since a `PUT` request replaces a resource and thus needs all the information, even for those we are not changing. We delete some API-specific information such as the `links` and `urls` sections from the returned data. With this done, we can then modify the property we want to change—in our case, the enabled status of our access rule, which we are setting to `False`. We then send a `PUT` request to the API. If the request is successful, we get the changed access rule object back.

Deleting access rules

The final capability we are looking for is how we can automate housekeeping tasks such as finding and deleting outdated rules. We have already covered finding a rule by retrieving all of the rules and iterating over their details in previous chapters, so this recipe will show you how you can delete an existing access rule.

Getting ready

Open your code editor and start by creating a file called `delete_access_rules.py`. Next, navigate in your terminal to the same directory in which you just created the `delete_access_rules.py` file.

How to do it...

Follow these steps to delete an existing access rule for your access policy:

1. Import the required `requests` library as well as the `json` and `sys` modules from the standard library:

```
import requests
import json
import sys
from requests.auth import HTTPBasicAuth
```

2. Specify the required variables such as your username, password, domain ID, and the ID of the policy, as well as the ID of the rule we want to delete:

```
base_url = "https://fmcrestapisandbox.cisco.com"

username = "<Insert username here>"
password = "<Insert password here>"
```

```
domain = "<Insert domain id here>"
policy_id = "<Insert your policy id here>"
rule_id = "<Insert your rule id here>"
```

3. Define an authentication function and use this function to obtain an authenticated session based on your username and password combination. Please have a look at the *How to do it…* and *How it works…* sections of the *Authenticating against the FMC REST API* recipe:

```
def get_authenticated_session(user, password, base_url):
    auth_url = f"{base_url}/api/fmc_platform/v1/auth/
generatetoken"
    auth = HTTPBasicAuth(user, password)

    resp = requests.post(auth_url, auth=auth,
verify=False)

    s = requests.Session()
    if resp.ok:
        s.headers.update({
            'X-auth-access-token': resp.headers.get('X-
auth-access-token')
        })
        s.verify = False
        print("Authenticated successfully!")
        domains = json.loads(resp.headers.get('DOMAINS'))

        return s, domains
    else:
        print(f"Failed to authenticate. Response code:
{resp.status_code}")
        sys.exit(-1)
sess, domains = get_authenticated_session(username,
password, base_url)
```

4. Next, we want to define the URL of our rule using the domain ID, policy ID, and rule ID. With the URL specified, we can send a DELETE request to the API server:

```
url = f"{base_url}/api/fmc_config/v1/domain/{domain}/policy/accesspolicies/{policy_id}/accessrules/{rule_id}"

resp = sess.delete(url)
```

5. Finally, we need to check if our DELETE request was successful:

```
if resp.ok:
    print(f"Deleted rule with id {rule_id}")
else:
    print(f"Failed to delete rule with id {rule_id}. Status code: {resp.status_code}")
```

6. To run this script, go to your terminal and execute the following command:

```
python3 delete_access_rule.py
```

> **Important note**
>
> Before executing this script, be aware that this will delete the access rule.

Your output should look like this:

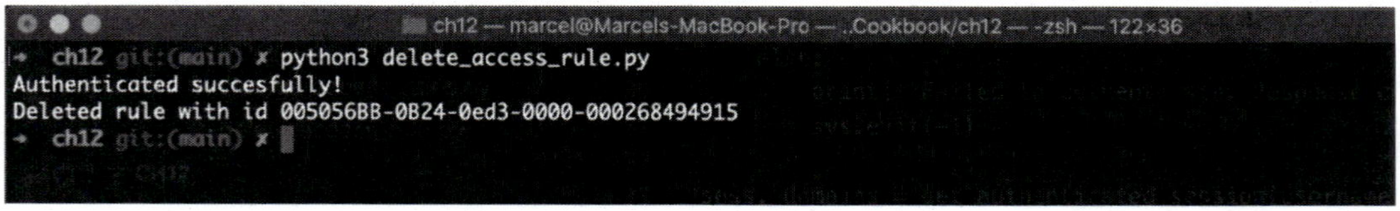

Figure 12.11 – Output of our access rule deletion script

The ID of your access rule will differ.

How it works...

We start by importing our required libraries, and then define our connection details such as the username/password combination to use for authentication, the base URL of our FMC instance, the domain under which we are operating, the policy for which our access rule is defined, and the ID of the rule itself. Next, we define our authentication function. For a more detailed explanation of how the authentication process works, please refer to the *How to do it…* and *How it works…* sections of the *Authenticating against the FMC REST API* recipe.

Next, we define the URL of the access rule we want to delete using the three IDs of the domain, policy, and rule. We then send a `DELETE` request to the API and check for a successful return of our request.

Other Books You May Enjoy

If you enjoyed this book, you may be interested in these other books by Packt:

Mastering Python for Networking and Security - Second Edition

José Manuel Ortega

ISBN: 978-1-83921-716-6

- Create scripts in Python to automate security and pentesting tasks
- Explore Python programming tools that are used in network security processes
- Automate tasks such as analyzing and extracting information from servers
- Understand how to detect server vulnerabilities and analyze security modules
- Discover ways to connect to and get information from the Tor network
- Focus on how to extract information with Python forensics tools

Azure Networking Cookbook - Second Edition

Mustafa Toroman

ISBN: 978-1-80056-375-9

- Get to grips with building Azure networking services
- Understand how to create and work on hybrid connections
- Configure and manage Azure networking services
- Explore ways to design high availability network solutions in Azure
- Discover how to monitor and troubleshoot Azure network resources
- Work with different methods to connect local networks to Azure virtual networks

Packt is searching for authors like you

If you're interested in becoming an author for Packt, please visit `authors.packtpub.com` and apply today. We have worked with thousands of developers and tech professionals, just like you, to help them share their insight with the global tech community. You can make a general application, apply for a specific hot topic that we are recruiting an author for, or submit your own idea.

Share Your Thoughts

Now you've finished *Python Network Programming Techniques*, we'd love to hear your thoughts! Scan the QR code below to go straight to the Amazon review page for this book and share your feedback or leave a review on the site that you purchased it from.

`https://packt.link/r/1838646639`

Your review is important to us and the tech community and will help us make sure we're delivering excellent quality content.

Index

Symbols

A

B

I

J

K

L

M

N

Q

R

S

T

U

V

W

Y

Made in United States
North Haven, CT
05 November 2026

21778844R10187